S I G N A L S

The Science of Telecommunications

John R. Pierce

A. Michael Noll

SCIENTIFIC
AMERICAN
LIBRARY

A division of HPHLP
New York

Library of Congress Cataloging-in-Publication Data

Pierce, John Robinson, 1910–
    Signals: the science of telecommunications/John R. Pierce, A. Michael Noll.
      p. cm.
    Includes bibliographical references.
    ISBN 0-7167-5026-0
    1. Telecommunication. I. Noll, A. Michael. II. Title.
TK5101.P54 1990   89-70207
    621.382—dc20   CIP
ISSN 1040-3213-5026-0

Printed in the United States of America.

Scientific American Library
A Division of HPHLP
New York

Distributed by W. H. Freeman and Company
41 Madison Avenue, New York, New York 10010 and
20 Beaumont Street, Oxford OXI 2NQ, England

1 2 3 4 5 6 7 8 9 0 KP 9 9 8 7 6 5 4 3 2 1 0

This book is number 32 of a series.

Contents

To William O. Baker, who saw the light and showed it to us both.

Preface

This has been a century of telecommunication. We have come to talk warmly and directly over the telephone rather than writing considered letters and fretting for a reply. Facsimile serves us better than first class or express mail. Through radio, compact discs, and television, we experience directly and movingly people and events far from us in space or in time. Through telecommunication, our world has expanded, and we with it.

In *Signals*, I set out to describe the science and technology involved in this revolution of telecommunication. I wanted to show how even though the laws of nature constrain what we *can* do, knowledge of those same laws has shown us how to accomplish far more than fantacists had dreamed. I wanted to show how researchers and inventors harnessed the forces of nature that cannot be circumvented.

This alone proved not enough. I found that things essential to the understanding of telecommunication were missing: people and their institutions. How could I tell this story without discussing such pioneers as Morse, Marconi, Bell, and Vail? Without Bell Laboratories and the Bell System? Without the government agencies that regulate communication in the United States and operate it in other countries? Beyond this, some technologically feasible communication services fail in the marketplace and others succeed. Consumers are an essential part of the story of telecommunication.

I, too, was missing from the book. I have met many of the brilliant technologists and scientists whom I mention, and some I know very well. Because I worked at Bell Laboratories, readers will see various aspects of telecommunication most clearly and in most detail in telephony, but the technology of telephony is basic to other areas of communication.

This book is a drastically revised version of an earlier edition of *Signals*, published nearly ten years ago by W. H. Freeman and Company, the publishers of the Scientific American Library series. A lot has happened in telecommunication during the past decade. As I foresaw, the world of communication has gone digital as fiber-optic transmission systems and digital switching became better, cheaper, and ubiquitous. Further, the old Bell System is no more. On January 1, 1984, AT&T divested the Bell operating companies. Competition became the guiding principle for telecommunication in the United States. These and other, rapid changes are now part of *Signals*.

This book could never have been written without the help and inspiration of a number of people. I am particularly indebted to many good friends at Bell Laboratories, most retired, for their help and inspiration over the years, especially Kenneth G. McKay, C. Chapin Cutler, C. H. Elmendorff III, Edward E. David, Jr., Peter B. Denes, Max V. Mathews, Manfred R. Schroeder, and M. Mohan Sondhi. To William O. Baker I am eternally grateful for all that he taught me and encouraged me to do. Dean Gillette, now of Claremont College, and Robert W. Lucky, still at AT&T Bell Laboratories, were kind enough to read and comment extensively on the manuscript of the present book. Dr. Hiroshi Inose of the National Center for Science Information System in Tokyo was kind enough to comment on the switching chapter. Richard Q. Hofacker, Jr., of AT&T Bell Laboratories provided much data and several illustrations as did many friends and associates at AT&T, the FCC, and other institutions.

My editor at the Scientific American Library, Susan Moran, helped clarify many of the ideas presented in the book, and her gentle prodding for even more clarity challenged me often to rewrite material. She has been a valuable contributor to *Signals*.

A few words need to be said about my use of *I*. In this book, *I* is used as a collective pronoun meaning both John R. Pierce and A. Michael Noll. We are quite different people, but we have managed to be like-minded about the technology, history, and impact of the telephone. In the few places where a specific identification of *I* would clarify matters, the identification is made.

I have an enduring love of telecommunication and relish the challenge of explaining its technology to as many readers as possible. I hope you enjoy reading *Signals* as much as I have enjoyed telling the story.

John R. Pierce
A. Michael Noll

January 1, 1990

Signals

1

·– – – –

A Grand Dream in a Real World

I love the story of the son who called his mother and invited her to fly across the country to attend her granddaughter's wedding. "Me, get into one of those planes?" she responded. "Never! I'll sit by my fire and watch television, as the good Lord intended."

We have unwittingly come to regard the telephone and television as parts of our natural environment, or, perhaps, as parts of ourselves. Telephoning is just talking to someone. Television is watching events happening elsewhere, as if we were looking at a scene just outside our door.

Our Telephone, but Their Television

Through the window of television we see what is happening in a host of regions that have become a part of our immediate world.

Alexander Graham Bell in his study. Versatility and insatiable curiosity led this teacher of the deaf into aeronautics, sheep breeding, and the presidency of the National Geographic Society. We know him because he invented the telephone and gave his name to the Bell System.

The telephone is ours from the time we learn to talk for as long as we can hear—and then some.

Hundreds of millions of people watched Neil Armstrong take his first small step on the surface of the moon. Since then we have seen a *Viking* spacecraft scooping soil from the surface of Mars, and we have admired the astonishing pictures of Jupiter, Saturn, Uranus, and Neptune, and of new rings and moons that the *Voyager* spacecraft sent back. Sadly, we have also witnessed the tragic *Challenger* disaster.

The telephone is fundamentally different from television. Not because it goes only where people go. Not because we hear but do not see. It is different because we use the telephone as an extension of ourselves. We use it to speak to anyone whom we choose to call, at any hour of the day or, alas for the sleeper, at any hour of the night. We speak *our* thoughts or feelings. In the words of the song, "a sigh is still a sigh," even though it emerges from the plastic earpiece of a telephone set.

A sigh may sadden a heart a continent away, but to do so it must somehow get into the telephone network and out again, passing through circuits owned by various enterprises, but serving you and me. The telephone network cannot transmit pleasure or anguish. Joy and pain lie in the human heart. The telephone transmits an electric signal that makes it possible to re-create the sound of *our* sighs at the far end of a circuit. The sigh itself, your sigh or my sigh, must convey the emotions that lie behind it.

The electric signals of television convey to us the sighs, shouts, and exhortations of others. Like radio and the press, television is from the few to the many. It brings us what someone else thinks we should know, or buy, or feel. We own our television sets, yet in some deep way television is *theirs*. We own none or only a small part of the telephone network, the telephone set and the wires in our home, yet the complex international telephone network is *ours*, to use as we will.

The distinction between mass communication—the press, radio, and television—and the person-to-person communication of the mails and the telephone has been appreciated by pragmatists as well as philosophers. Leon Trotsky wrote of Joseph Stalin's opposition to a proposal for an effective telephone system in the USSR. Stalin said that no greater instrument for counterrevolution and conspiracy could be imagined. In her book *The Telephone in a Changing World*, Marion May Dilts noted that Adolf Hitler, who fully exploited the mass media, stopped telephone development in Germany by imposing large taxes. The Supreme Court of the United States has ruled out special taxes on newspapers as unconstitutional, yet allows taxes on telephone service. Govern-

ments distinguish sharply between *our* person-to-person communication by telephone and the few-to-many communication of the press, radio, and television.

Where Technology Reigns

Providing person-to-person service for many people presents great technological difficulties. Compare radio and television with telephony. All a broadcaster needs is transmitting equipment and a standard studio. Out there are a lot of people who have bought receiving sets. The broadcaster sends the same programs to everyone. Programs are transmitted from station to station through the facilities of companies or organizations whose chief business is telephony and data transmission.

   The technology of telephony is inherently far more complicated than that of broadcasting, primarily because *any* subscriber must be able to reach *any other* subscriber. Broadcasting each message won't do. A town crier can't shout around the world, and if he could, we wouldn't want our conversations made public. To reach

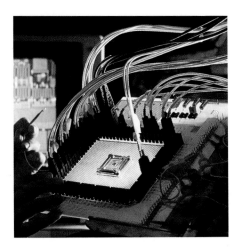

The advance of telephony has come through the exploration of plausible ways to do new things, or to do old things better. This neural network chip developed by Bellcore in 1988 is the first such chip to emulate the structure and learning behavior of the human nervous system, enabling the chip to "think" and make decisions. Proposed applications have included the recognition of written characters and of spoken words.

the ear of another, the subscriber keys the location (the phone number) to be reached into a computer, which searches out among a huge array of existing communication equipment a possible path between caller and called, and establishes that path. The computer also notes any special billing charges—or, if the number dialed is not in service, the computer explains this to the caller.

Telephone companies have not only a more difficult task than broadcasters, but they are held to much higher standards of service. If a radio station goes dead, the listener may grumble. If telephone service is disrupted, the user will scream. The user pays for service and expects it to work. And, in most countries, there is a long tradition of reliable telephone service. More reliable by far than electric power.

This is remarkable, because the telephone system is by far the most complicated machine ever constructed by human beings, and today different parts are owned and maintained by different companies. You can pick up a telephone and dial the number of a friend in Los Angeles or Hawaii—or even in Sri Lanka. The call will be completed without human intervention and terminated when you hang up. Machines will prepare itemized bills that are sent to you, bills for local service and bills for long-distance calls.

It is not surprising that with its emphasis on technology, telephony also differs sharply from mass communication in being dominated by engineers. The late Ithiel de Sola Pool, professor of political science at the Massachusetts Institute of Technology and editor of *The Social Impact of the Telephone*, wrote in that book: "Today phone systems are usually dominated by engineers, but in broadcasting organizations, engineers play a rather lowly role. The top positions in commercial systems are held by people from either the programming or marketing side, and in government by civil servants."

This may not be true in the United States to the extreme degree that it was before the Bell System was dismantled in 1984. Yet, because the telephone system deals with service, not content, and because the technological difficulties remain, enginers will continue to play a strong part in telephone organizations.

Cost and Value, Complexity and Simplicity

A successful technology must not only work, it must be affordable to both the users and the providers. The customer's equipment, the telephone, is an almost negligible fraction of the telephone plant.

The Telephone System

Behind every telephone is an elaborate system made up of various kinds of transmission media that carry voice and data signals from place to place and switching systems that connect the circuits linking together two parties.

Telephones in the home are connected to the network by a modular plug. All the telephones in the home transmit their electric signals over pairs of wires that terminate at the protector. The protector functions as a fuse or circuit breaker to prevent large electric currents and voltages from damaging the telephone instruments.

From the protector, the telephone signals are carried over a twisted pair of wires, called the local loop, to the central office.

Many local loops share cables carrying as many as thousands of pairs of wire. All these wire pairs are connected to a switching system located in the central office.

This local switching system decodes the digits dialed or keyed by the caller. If the calling party is also served by the same local switching system, the final connection will be made there. If not, other transmission paths, called trunks, carry the signals to other local offices.

In the case of a long-distance call, trunks connect the local office to the nearest switching system located in a long-distance network. A long-distance network consists of many switching systems and transmission circuits across the country

and, in some cases, linking continents.

The United States has been divided into a number of local access and transport areas (LATAs). The local telephone companies provide intra-LATA service only. Service from one LATA to another (inter-LATA service) is provided by a variety of long-distance common carriers. Competition has come to inter-LATA service, but the provision of intra-LATA local service is still nearly the exclusive domain of the local telephone company. One possible reason for this is that the provision of local service is labor intensive and does not enjoy the large economies of scale of long-distance service.

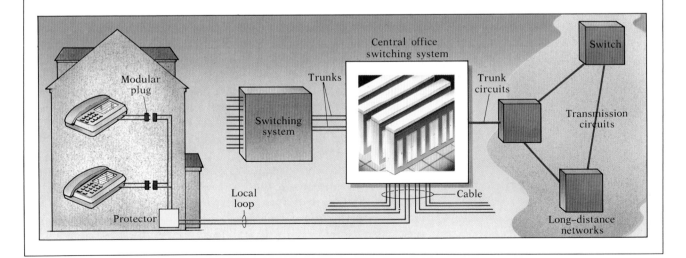

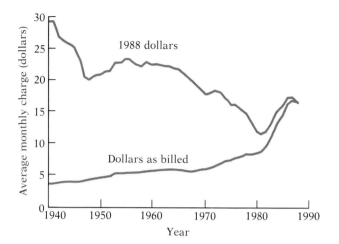

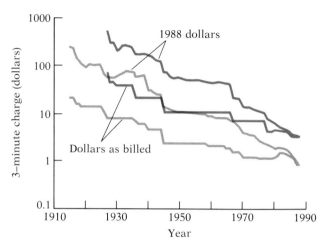

Left: The graph shows the average monthly rate for U.S. local single-party service with unlimited calling, including taxes but excluding telephone instruments. Rates have gone down steadily, except for increases that started a few years before the divestiture of the local Bell companies from AT&T. Right: Technological innovation, especially today's optical fiber, has been most dramatic in lowering steadily the charges for calling coast to coast and other long distances. The graph shows the station-to-station day charges for a three-minute call from New York City to Los Angeles (orange) and from New York City to London (blue) at the actual billed rate, and converted to 1988 constant dollars.

Most of the cost and complexity lie hidden in wires to central offices, in switching equipment, and in transmission facilities between offices. Because of the problem of connecting subscribers, an inherent diseconomy of scale exists when telephone service is provided to many people. An early telephone-company manager is quoted as saying that all he had to do was to get enough subscribers and the company would go broke.

Yet, there is a sort of economic paradox. The *value* of having a telephone rises with the number of people you can talk to, the number of people who have phones. An early and powerful force in telephony was the drive to increase its value by making it a universal service. In the United States and some other countries, in endeavoring to get and keep subscribers, part of the cost of local telephone service was paid for by long-distance revenues. This force is reflected today in the United States in the connection charges that long-distance carriers pay to the phone companies for the privilege of receiving business from the host of subscribers.

But the cost of serving many subscribers is high. Connecting each telephone directly to every other telephone of the 200 million in the United States alone would be ridiculously expensive. It is far more efficient to connect each telephone to a central office by individual wires and make the connections between the individual wires at that central office. Still, the cost per subscriber must increase with the number of subscribers. The cost increase per new subscriber is not only the cost of the wires to the subscriber's location, but also a share in the ever-increasing complexity of the

CHAPTER 1

switching systems that are necessary to reach many millions of subscribers, today scattered over all of the countries of the world. Only high technology combined with economies of scale in research, design, manufacture, and operation can keep the cost of telephony down as service expands. What degree of common ownership is optimal we may never know, but if service were provided by a host of local and truly independent companies, few could afford it.

The wonder is that anything as complicated as telephony could become the universal service that it has become in many countries. Indeed, it is still a wonder to me.

The two authors of this book worked for Bell Laboratories in research for a combined total of 50 years, about half of the telephone's history. Only gradually did we come to appreciate the necessity as well as the fascination of technical progress in communication. It has made universal and cheap telephone service possible and has maintained such service despite the system's increasing complexity, rising wages, rising costs of materials, political obstacles and opposition, and finally, the breakup of the national telecommunications system into a number of companies that both cooperate and compete.

The Growth of Telephone Technology

Today the telephone is indispensable to our everyday lives as well as to business. We do part of our shopping by telephone. We consult doctors, get information, and arrange meetings over the telephone. We pour out our hearts to those we cannot meet. We use the telephone to call people whom we would not or could not go to see in person, and we sometimes say things over the telephone that we would not say face to face.

Without the telephone, skyscrapers would have been unworkable; the elevators would have been jammed with messengers. When I, John Pierce, first went to Bell Laboratories in 1936, the old oak desk that was given to me was equipped with a white push button connected to nothing. An employee of an earlier, pre-Depression generation was able to push that button and summon a messenger to go to the stockroom, to cash a check, or to run some other errand.

Before the telephone, owners or managers of enterprises had their offices at their plant, where they could supervise operations. With the coming of the telephone, head offices could be centralized

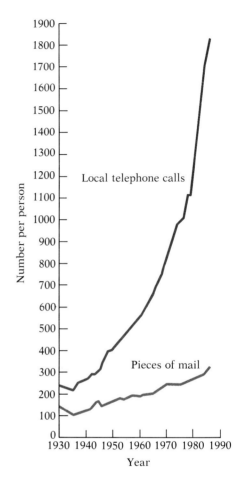

Over the past 60 years, telephone calls have far outdistanced letters as a means of person-to-person communication in the United States. The number of telephone calls per person per year is nearly six times the number of first-class letters and airmail. Unless society changes fundamentally, a labor-intensive service such as mail delivery will continue to decline in comparison with an automated service such as the telephone.

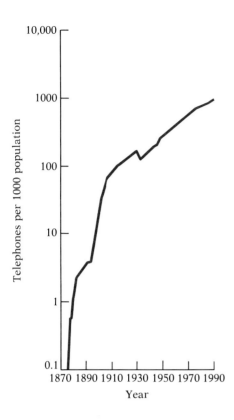

From the year the telephone was invented, its use grew rapidly in the United States, with the sharpest increase in the rate of growth occurring in the mid-1890s.

in New York and other large cities. Today, many of the operations of insurance and other service companies have moved to communities where wages and living costs are low. Telephony and data communication have made decentralization possible.

The early development of telephone service in the United States stemmed from a clear and early idea of the telephone as a universal person-to-person service. In 1878, two years after his invention of the telephone, Alexander Graham Bell stated that idea very clearly. In an address to a group of London capitalists associated with the Telephone Company in England, he described switched telephone systems and said, "I believe in the future wires will unite the head offices of the Telephone Company in different cities, and a man in one part of the country may communicate by word of mouth with a different place."

Works as well as faith were necessary for the realization of this dream. In 1878 Theodore Vail, then a young mail superintendent for a railroad company, joined the Bell Telephone Company. Although Vail left the company in 1887, he later served as president of its successor, the American Telephone and Telegraph Company (AT&T) from 1907 to 1919.

Vail shared Bell's vision. Vail's favorite slogan was "One policy, one system, universal service." In 1879 he wrote to one of his staff, "Tell our agents that we have a proposition on foot to connect the different cities for the purpose of personal communication, and in other ways to organize a grand telephone system."

Those were bold and perhaps foolhardy words to utter before any long-distance communication had been demonstrated, but they were spoken by a man of great business acumen who had faith in the power of science and technology. That faith proved to be fully justified.

From the year the telephone was invented its use grew rapidly. Only two years after the first exchange was set up in 1878 at New Haven, Connecticut, many major European cities were offering telephone service. But at first service in most other countries lagged behind that in the United States. All of Europe had 97,000 subscribers in 1887 compared to 150,000 in the United States. Some countries did not see clearly the nature and promise of phone service; in Hungary it was first promoted as a wired broadcast service into the home. Eventually the unique nature of telephony became apparent, and service grew in all technologically advanced nations, although at different rates and to different degrees of usage.

CHAPTER 1

Alexander Graham Bell

"Leave the beaten path and dive into the woods," Bell said. "You are certain to find something interesting." Of Bell, the great physicist James Clerk Maxwell said, "Now, Prof. Graham Bell, the inventor of the telephone, is not an electrician who has found out how to make a tin plate speak, but a speaker who, to gain his private ends, has become an electrician." Bell was the son of Alexander Melville Bell, a distinguished student of speech. Alexander Graham Bell worked with his father in London, then moved to Canada in 1870 and to the United States in 1871. He taught speech to the deaf and became a professor of vocal physiology at Boston University in 1873. The idea of sending the sound waves of speech by means of an electric current came to Bell in 1874, and he devoted the next two years to developing the telephone, which, as we have seen, he patented in 1876. The word *telephone* comes from the Greek words meaning "far" and "sound." Bell had an astonishingly broad and accurate view of the place his invention would take in society, and in 1878 he predicted widespread local service of the sort we have today. In 1877 Bell married Mabel Hubbard, a deaf student whose father had helped support Bell's experiments. His later inventions included the photophone, which transmitted sounds by means of light waves, and the audiometer, a device still used to test hearing. In the latter part of his life Bell became interested in flight (the tetrahedral kite was his invention) and in the breeding of sheep that would bear twins. He died in Baddeck, Nova Scotia, in 1922.

Did Bell invent the telephone? The Supreme Court upheld Bell's patent in a contest with Elisha Gray, who was backed by Western Union. I regard Bell as the inventor.

This 1890s pay phone was intended to give more people access to telephones. Cheaper home service won out in the end, but pay phones are still with us.

The United States experienced its sharpest rate of growth in the mid-1890s. After the expiration of Bell's original patent in 1893, many competing companies were formed, some of them providing service where none had been available. In Europe, the progression was the opposite. At first most nations granted licences to competing private companies for different cities. But by the 1890s many governments were taking control of the private telephone networks. The initiation of commercial long-distance service fostered the expansion of telephony near the turn of the century.

Early telephone rates were not cheap. The Bell Telephone Company tried various expedients in an effort to make telephone service more generally available, including metered service and the pay phone, which was first introduced in Springfield, Massachusetts, in 1893. In the end, it was through continuous and determined progress in science and technology that the telephone became a universal service.

The strong influence of technology is demonstrated by the rapid rise in overseas calls. Commercial overseas telephony did not exist until radio service was established in 1926, and consistent high-quality service came only with the first transatlantic telephone cable in 1956. Since the initiation of commercial telephone service by satellite in 1965, the number of nations that can be reached by telephone has increased greatly. For service to highly developed areas such as Europe and Japan, optical fiber cables are better and cheaper than satellites, but satellites are invaluable in providing telephone and other communication circuits to small or remote nations, and within the island nation of Indonesia.

From the start, the various Bell companies undertook the assiduous pursuit of technical advances. Their pursuit of science, technology, and invention led to the formation of the Bell Telephone Laboratories in 1925. Bell Laboratories, now detached from the companies that provide local telephone service, continues (with some modifications) as AT&T Bell Laboratories. Most major countries established research laboratories as well.

I think that the part that rapid technical progress plays in telephone service is largely hidden from the everyday user of the telephone—that seemingly simple and inexpensive device with which elaborate data services find it hard to compete. Superficially, telephony seems to change slowly. Such innovations as new types of phones, automatic switching, transoceanic service, and direct long-distance dialing have been spaced years apart.

The subscriber does not see most changes in the telephone plant. The provision of an ever more extensive and complicated

The Initiation of the Earliest Long-Distance Routes

DATE	CITIES
1881	Boston to Salem, MA
1884	New York, NY to Boston, MA
1892	New York, NY to Chicago, IL
1893	Boston, MA to Chicago, IL
1893	New York, NY to Cincinnati, OH
1895	Chicago, IL to Nashville, TN
1896	Kansas City, MO to Omaha, NE
1896	New York, NY to St. Louis, MO
1897	New York, NY to Charleston, SC
1897	New York, NY to Minneapolis, MN
1897	New York, NY to Norfolk, VA
1898	New York, NY to Kansas City, MO

service in the face of increased costs of labor and materials has led to a continual search for new materials, new devices, new techniques, and new methods. Plastics have replaced the wood, metal, and hard rubber of the old telephone set and the expensive lead of cable sheathing. The latest cables transmit signals through optical fibers rather than by wires. Highly purified semiconductors have made possible integrated circuits and light-emitting diodes and lasers.

The first telephones transmitted one signal at a time on single wires with a ground return (using the earth as a second wire). Now, tens of thousands of conversations travel simultaneously over optical fibers. Although microwave radio has long sent calls across the continent, it can now send signals all over the world via communication satellites.

Smaller and less costly transistors, embedded in integrated circuit chips, have replaced vacuum tubes in the amplification and control of signals. Switching systems have been transformed in a sequence of steps since their very beginning in 1878. Originally performed manually by telephone operators, switching is now controlled completely by computers.

Only through decreasing costs to customers in the 1920s could phone service gain wide-spread acceptance, both in the home and in businesses.

The Laws of God, the Laws of Man

The achievement of universal telephone service has been a unique technological challenge. Those who provide it have had to integrate all sorts of new technologies into a stable and reliable system, a system so standardized and automated that by pushing buttons on a telephone a subscriber can cause a phone to ring in a far, far land, and can converse with a friend there. It is through technological advances that such service has been provided, and made cheap enough so that most of us can afford it.

I came to realize during my years at Bell Laboratories that the challenges that have faced and still face telephony go far beyond the challenges of the laws of nature. That realization was driven home through my interest in harnessing the then-new technologies of space exploration in the service of telephony.

At one time the chief concern of one of the authors, John Pierce, was transoceanic communication by means of satellites. The first balloon satellite was my idea. Rudi Kompfner, other co-workers, and I, along with Bill Pickering, then head of the Jet Propulsion Laboratory at the California Institute of Technology, persuaded NASA to fund the project. In 1960, NASA launched the *Echo* balloon satellite, which in demonstrating transcontinental voice transmission used the East-Coast transmitting and receiving terminal built by people in my division at Bell Laboratories.

The success of *Echo* led AT&T to launch *Telstar* in 1962. *Telstar* was the first satellite to transmit telephone and television signals across the Atlantic. Ironically, that same year the Communications Satellite Act, which created the Communications Satellite Corporation (Comsat), legislated the Bell System out of the international communication satellite business. The Bell System, in its pursuit of technological aims, continued to suffer setbacks through government regulation. Thus, in order to favor satellite communication, the FCC tried to maintain an even mix of traffic between satellite and cable circuits, regardless of cost. For this reason, an advanced underseas cable, the TAT-7, was delayed from 1981 to 1983. The FCC also held up permission to construct an advanced mobile communication system from 1971 to 1977, and only in 1981 was full-scale commercial operation approved.

The English poet A. E. Housman advises,

> *The laws of God, the laws of man,*
> *Let him keep who will and can*

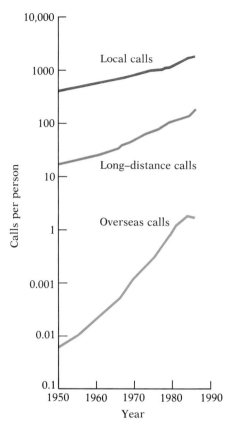

The number of long-distance calls placed each year from the United States rises more rapidly than the number of local calls, and the number of overseas calls increases more rapidly than the number of long-distance calls. The trend reflects both the falling cost of long-distance service and changing patterns of life: the tendency of families to disperse, keeping in touch by phone, and commercial, scientific, and cultural relations that are increasingly nationwide and international.

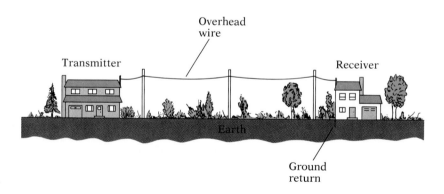

The simplest telephone circuit consists of a single overhead wire connecting a transmitter to a receiver, and a ground return through the earth. The earth functions as a second wire and completes the telephone circuit.

The Echo *balloon satellite transmitted the voice of President Eisenhower across the American continent and alerted the world to the potentialities of communication satellites.*

My work was concerned with the natural laws of God. In the end I came to realize that while the sort of communication that *can* be provided depends on technological feasibility, the sort of communication that *will* be provided depends on the laws of man. Technology persists as a basic theme of telephony, but telephony cannot be understood without taking into account the laws of man. I will have more to say on this topic in Chapter 10.

What Lies Ahead

The intriguing problems of electrical communication in general and of the telephone in particular are the chief concern of this book. I hope to tell how and why telephony has become the subtlest as well as the most widespread form of electrical communication.

The human aspect of communication is common to our everyday experience. But if we are to understand and appreciate telephony, we must stretch our minds toward the underlying principles of nature, the laws of God in my metaphor, toward concepts of universality and of great utility. These concepts are the encoding of sound and sight into electric signals; the nature of signals and communication channels; the theory of information, which tells us how we can quantify sources of signals and the channels that transmit them; and the practical and subtle art of modulation, through which signals can be represented appropriately and combined for transmission over a single medium. We must also think of the physics of communication—the nature of the electromagnetic waves of radio and light and the physical nature of the vacuum tubes, transistors, integrated circuits, and lasers that produce waves and process signals.

Once we have some sense of the processes and the physical nature of communication and of the devices through which we communicate, we can gain some sense of a communication system in which signals are generated, transmitted, and switched. We can understand how new developments in technology can increase the impact that communication already has on our society. The final outcome of any new technology depends on human reactions. Science and technology can point to plausible paths, but consumer choice determines what will succeed or fail.

Knowledge is hard earned. But, without knowledge, we can do no more than fantasize, which is childishly easy. The knowledge that can take us beyond fantasy requires an exercise of the mind, an exercise that can be as invigorating as exercise of the body.

The advance of optic-fiber communication depends on the production of extremely pure semiconductor materials such as this single crystal of indium phosphide grown under high temperature and pressure.

2

The Beginnings of Electrical Communication

When Pedro II, emperor of Brazil, traveled to the United States in 1876, he visited the Boston School for the Deaf, where he met Alexander Graham Bell, a teacher of the deaf. A short time later, at the International Centennial Exposition in Philadelphia, Bell demonstrated his new telephone for the emperor. "My God," Dom Pedro is said to have exclaimed, "it speaks Portuguese!"

That story may or may not be true. The point is that because all people can talk in some language, they can all use the telephone without training. Only technological and economic success were required to make telephony universal. The users did not need to learn a new skill. They did not even have to know how to read or write.

In the beginning of telegraphy, a messenger brought good or bad news to your home. Today, telegrams are delivered by phone or mail.

Samuel F. B. Morse

Alexander Graham Bell was not an electrician by training; he was a teacher of the deaf. Morse was not an electrician by training either; he was a painter, and founder of the National Academy of Design. Like Bell, to whom the services of Thomas Watson were invaluable in making the telephone, Morse sought the help of men with skills not his in making the telegraph. Among these were Leonard Gale and Alfred Vail.

When I read Carleton Maybee's generally excellent life of Morse (*The American Leonardo, a Life of Samuel F. B. Morse*), I was puzzled by Maybee's almost apologetic account of what he saw as a lack of novelty in Morse's invention. Morse clearly was the first to have the idea of an electric telegraph operating on a single electric circuit that communicated by means of a code. A modification of his Morse code became an international standard. In an entirely different context, codes of sequential symbols for the representation of text had been devised long before, but Morse probably did not know of them. Other telegraphs had been devised: semaphor telegraphs, which may have given Morse a clue, and electric telegraphs that used several wires in order to make a needle point to a letter. Morse's telegraph was the right telegraph. It set an enduring pattern of encoded transmission over a single circuit that is followed today in sending text and data letter by letter (or symbol by symbol). Such evidence as there is indicates that Morse's associates *helped* him to realize his dream and to implement his particular ideas.

Samuel Finley Breese Morse was born in Charlestown (now a part of Boston) on April 27, 1791. He entered Yale University at the age of fourteen and graduated in 1810. Although he attended lectures on electricity while at Yale, after graduation he went to England to study art. He returned to the United States in 1815 and became well known as a portrait painter.

In 1832, while returning from Europe, he and a fellow passenger discussed the electromagnet, and

Morse conceived the idea of his telegraph. He made a working model about 1835, filed for a patent in 1838, and in 1844 inaugurated public service between Washington, D.C. and Baltimore with his famous message, "What hath God wrought?" Morse's code, shown on the left, exhibits high ingenuity because it assigns short codes to commonly used letters such as *e* and long codes to infrequently occurring letters such as *z*. All practical telegraphy stemmed from Morse's invention. By the time of his death in New York in 1872 Morse had received honors from all over the world.

A	. _	N	_ .
B	_ . . .	O	_ _ _
C	_ . _ .	P	. _ _ .
D	_ . .	Q	_ _ . _
E	.	R	. _ .
F	. . _ .	S	. . .
G	_ _ .	T	_
H	. . . .	U	. . _
I	. .	V	. . . _
J	. _ _ _	W	. _ _
K	_ . _	X	_ . . _
L	. _ . .	Y	_ . _ _
M	_ _	Z	_ _ . .

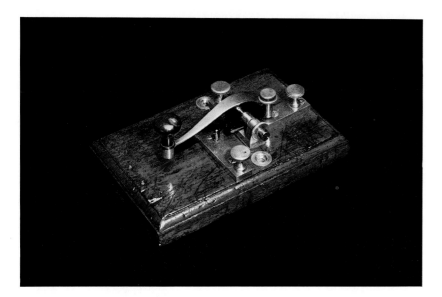

This telegraph key, much like some in use today, was used on the 1844 Washington to Baltimore telegraph line. In sending Morse code it proved easier simply to close and open an electric contact by pushing the key than to set type in a portrule and pull it through Morse's original transmitting device.

Telegraphy

All modes of communication have problems, but the problems differ for different modes. In Greek legend it is said that Cadmus, a Phoenician prince, taught the Greeks letters, but we have no record of his struggles in adapting a new mode of communication to their needs. The early history of Samuel F. B. Morse's telegraph shows how hard it is to gauge human abilities and to fit a novel form of communication to them—a struggle Bell was spared because he required only that a user be able to speak and hear.

When Morse cast about for a means of electrical communication and conceived the idea of his telegraph, he devised various awkward schemes to send text over a wire by combinations of dots and dashes. At first he sent numbers only; to decode the message, the receiver looked up numbered words in a code book. Later Morse used codes of dots and dashes to transmit text letter by letter, creating the famous Morse code. But, through several years of experimenting, he used an elaborate sort of typesetter to generate signals and a recording device to write out the dots and dashes at the receiving end. Operators soon found that they could send dots and dashes easily and rapidly by pressing a simple telegraph key similar to the one used today by amateur radio opeators. And they discovered that they could interpret the clicks of the receiving instrument—the sounder—faster than they could read recorded dots and dashes off paper. Thus, telegraphy changed in its very inception to accord with the aptitudes of human beings.

The telegraph key and sounder are technologically the simplest method for transmitting a message over a distance. One reason Morse's telegraph did not attain the universality of the telephone is that people had to learn a code in order to use it. To circumvent this obstacle, inventors devised printing telegraphs

Morse's original transmitting device (bottom) and original receiving device (top). The message was set in a sort of type depicting Morse code and placed in a stick called a portrule. An operator turned a crank (bottom right) in order to pull the portrule past an electric contact device that sent the message. Closing the contact caused current to flow through an electromagnet in the receiving device. The flow of current pulled down a pen, and the pen wrote the transmitted dots and dashes on a moving paper tape that was pulled along by a weight-and-pulley (upper right).

CHAPTER 2

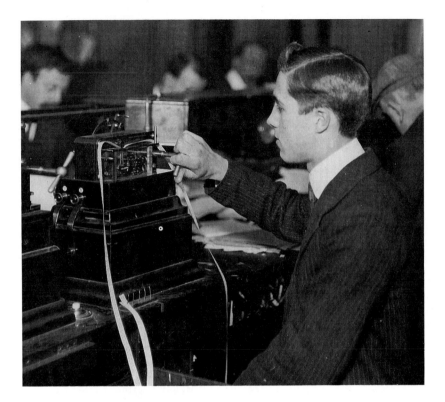

that could be operated by means of a keyboard, but this method was still too complex to be widely popular. Although today's personal computers are far more satisfactory than printing telegraphs, and can be and are used to send electronic mail, even these lack the universality of the telephone. Higher cost and absence of a single standard are factors, but for most purposes speaking is simply easier and better than writing.

Although the telephone works equally well for any language—English or Japanese or Chinese—keyboard devices do not. Messages in English can be transmitted simply and unambiguously by means of a keyboard. Each keystroke sends a short, simple code for a particular letter or symbol. But think of Japanese typewriters, which must print over a thousand characters.

Japanese can be written in phonetic characters called *kana*. Such text is ambiguous because Japanese contains many homo-

Because the Japanese use kanji, or Chinese-derived ideographic characters, in addition to the phonetic kana characters, the keyboard of a traditional Japanese typewriter had over 1000 characters, and typing was a demanding skill.

phones—different words with the same sound. (*Felt* hat and *felt* well are examples of ambiguity in English.) Further, Japanese names are in most cases represented by Chinese ideographic characters called *kanji*. In Japanese word processors, a word typed in kana evokes on the screen all possible kanji characters; the user makes a choice that holds until a new choice is made. Still, the encoding of text has been a major problem in Japanese, and in Chinese as well.

Facsimile

The difficulty of transmitting Japanese or Chinese text can be overcome by transmitting a black-and-white facsimile of handwritten or printed text. Indeed, the standardization and rapid growth of fax (facsimile) transmission over telephone connections has come in a large part from Japan, and has spread from there to the rest of the world. Transmitting text in this way is like transmitting a picture.

Although we read text by sight, it would be misleading to consider the transmission of text by means of a code for each letter and symbol, as in data transmission and computer mail, as visual communication. In such a representation of text, an *a* is an *a* and an *&* is an *&*, whatever it may look like. Script, hand printing, and different type styles and sizes look different, but in reading we interpret any recognizable shape of letter as the letter intended. Unless the type is very hard to read or very odd, we do not even notice the particular shapes of the letters. In transmitting alphabetical text we are usually content to transmit a code that identifies the letter and causes the creation of a shape that is readable.

Forms, contracts, and signatures pose problems that the encoding of characters cannot deal with. Mere readability doesn't suffice. In a real-estate transaction, for instance, offers should look right, and they should be signed. Fax makes it possible to send such documents, or an image of them, more quickly and conveniently than by mail. And one can send advertising copy, or marked-up text, or a host of other things that electronic mail deals with awkwardly or not at all.

In sending a picture, vastly greater amounts of information must be transmitted than sending a short code for each character. The need to transmit all of this extra information was once put forward as a serious handicap for fax. The telephone network has proved capable of handling this information at a small cost. Cheaper and more powerful integrated circuit chips and cheaper transmission made fax affordable. Facsimile terminals are now reliable, though they are still more complicated and more costly than telephone sets. Their advantages are so great that offices are "deprived" without them, and they are appearing in homes.

In the growth of facsimile we may see something of the early competition between the telephone and the telegraph. Telephone instruments had to perform a more complex function than telegraph instruments. They were used, and improved, because te-

---

Picture transmission is more difficult than the transmission of text.

Picture transmission is more difficult than the transmission of text.

Picture transmission is more difficult than the transmission of text.

Picture transmission is more difficult than the transmission of text.

Picture transmission is more difficult than the transmission of text.

Picture transmission is more difficult than the transmission of text.

Picture transmission is more difficult than the transmission of text.

The same sentence is shown in various styles of type. As text, all are equivalent and can be transmitted simply by sending a short code for each letter. To transmit such text as a picture displaying the complexities of each style of type is more difficult and costly, because it requires transmission of far more information.

lephony was "better"—more adaptable to more uses—than telegraphy.

Electricity

Bell was not an electrician, but a teacher who became an electrician in order to gain his ends. In understanding the problems of telephony one must, like Bell, become something of an electrician. I hope that readers of this book will come to understand electricity better than Bell himself did. In seeking to understand, it is best to proceed from simple illustrations to more complicated relations.

The idea of electric charge is fundamental. This idea came early in the investigation of electricity, when people found that a glass rod rubbed with fur or silk would attract light objects. The attraction was a manifestation of electric charge. We now know that objects can be charged positively or negatively, that like charges repel one another, and that unlike charges attract one another.

The atoms of all materials consist of a central nucleus with a positive charge surrounded by orbiting electrons of negative charge. In various ways one can add a few electrons to a material object and give it a negative charge, or subtract a few electrons and leave it with a net positive charge. The unit used to measure amount of charge is the *coulomb*. Electric charge in coulombs is commonly designated by the letter Q.

In communication we are less interested in electric charge itself than in the flow of electrons through metal wires or other conductors of electricity. The rate of flow of charge in coulombs per second is the electric current measured in *amperes*. Electric current in amperes is commonly designated by the letter I. Electric current can be measured by a device called an ammeter.

What causes an electric current to flow in a conductor? An *electromotive force* urges electrons through the conductor. This force is what a battery or an electric generator supplies. Electromotive force is measured in *volts*, and the electromotive force of a battery or generator is called a *voltage*. Voltage can be measured with a device called a voltmeter.

In causing an electric current to flow, a battery or generator supplies electric power. Power, measured in *watts*, is usually designated by the letter P. Power P is the product of the current in amperes times the voltage:

$$P = IV$$

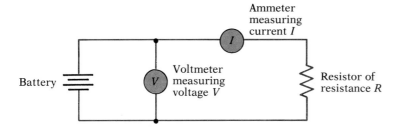

When a voltage V is applied to a conductor of resistance R, a current I will flow. Electric current is a flow of electricity, which may be likened to a flow of water. Electric voltage tends to make current flow; it may be likened to water pressure. Electric resistance is the measure of how much a particular conductor impedes the flow of current. According to Ohm's law, the relation among V, R, and I is V = IR, or I = V/R.

We are all familiar with electric power in terms of light bulbs and motors. Most of the powers we encounter in communication are much smaller. Further, we are very often concerned with the ratios of powers rather than with power itself.

An electric current flows into one terminal of an electric circuit and out of the other. An electric circuit can be a long pair of wires with some electric device, say, a telephone receiver, connected at the far end. If we connect the terminals of a battery of voltage *V* to the terminals of a circuit, a current of *I* amperes will flow through the circuit.

How great will *I* be? That depends on the electric *resistance* of the circuit, measured in *ohms*. The word *ohm* comes from the name of a German physicist, Georg Simon Ohm (1787–1854) who discovered Ohm's law

$$I = \frac{V}{R}$$

This law holds for almost all conductors.

Ohm's law is a linear relationship between the current *I* and the voltage *V* which causes the current to flow; if we plot current vs voltage we get a straight line of slope 1/*R*. Ohm's law holds for voltages and currents that don't change too rapidly. Here we will assume that current and voltage vary with time in exactly the same way, in accord with Ohm's law.

It is interesting and important to note that by using Ohm's law we find that electric power is proportional to the square of the voltage, or to the square of the current:

$$P = IV \qquad P = I^2R \qquad P = \frac{V^2}{R}$$

Sound Waves

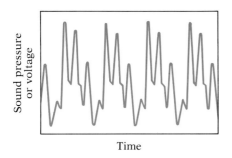

The telephone signal is a direct electric representation of the speech waveform. It is an electric voltage (or current) that varies with time in the same way that the pressure of the sound wave of speech varies with time as that wave travels through the air. The transmission of the direct electric analog of a voice signal is called analog transmission.

In the early days of telephony, the human voice was transmitted by means of electric current and voltage using equipment that was physically little if any more complex than the telegraph key and sounder. The quality of the transmitted voice was poor and the sound was faint. But users were astonished and pleased that they could catch what was said, even with difficulty. The scheme or idea of the telephone was simple; the early apparatus used was crude and imperfect, but for the user, *it worked.*

Telephony starts with a sound wave that is produced in the air by the act of speaking. The idea of a wave is somewhat tricky. When we throw a stone into a pond, ripples travel out in circles. But, the water doesn't bodily flow in the direction in which the ripples travel. What travels is a disturbance in the height of the water; in the wave this disturbance is continually passed on from one part of the water to the next. As the wave passes, the height of the water fluctuates with time in an undulatory (as opposed to an abrupt up *or* down) fashion.

We will encounter waves repeatedly in later chapters of this book. Waves can be traveling patterns of voltage or current, or of electric or magnetic fields. A sound wave is a traveling pattern in which the air pressure increases and decreases rapidly with time. The amount by which the air pressure increases and decreases is the *amplitude* of the wave. The power of the sound wave in watts per square meter is proportional to the square of the amplitude. Waves of greater power sound louder. How often the pressure of a smooth, simple sound wave rises or falls in a second is the frequency of the wave in hertz. (The hertz, abbreviated Hz, is a unit of frequency equal to one cycle per second. It is named in honor of Heinrich Hertz, who first produced electromagnetic waves. More will be said about him in Chapter 7.)

When we speak into a telephone, the pressure of the sound wave produces on a wire a corresponding wave of voltage and current. This electric wave of voltage and current travels along the wire. At the far end of the wire the sound wave is recreated from the electric wave.

A transmitter in the telephone handset turns a sound wave into an electric signal. In doing so the transmitter acts as an acoustic-to-electric *transducer*. A transducer is a device that converts variations in one physical quantity (as, in sound pressure) into corresponding variations in another physical quantity (as, in voltage). In turning the electric signal back into a sound wave at the far end of the wire, the receiver in the telephone handset acts as an electric-to-acoustic transducer.

In Bell's first telephone the reproduced sound wave was so faint that it was barely intelligible. The acoustic-to-electric transducer at the sending end was inefficient, so the electric power produced was less than the power of the original sound wave. The electric-to-acoustic transducer at the receiving end was also inefficient and produced an acoustic signal less powerful than the electric signal on the line, and much less powerful than the original acoustic signal.

But even in modern telephones, the re-created voice signal in the telephone receiver is never a perfect reproduction of the sound wave at the telephone transmitter. In what ways does the reproduced speech differ from the original speech?

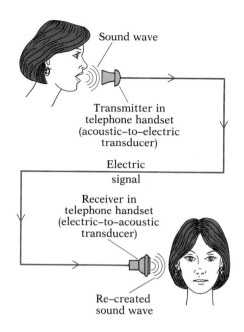

Telephone message transmission begins with the sound wave produced when a person speaks into the transmitter. An acoustic-to-electric transducer converts the fluctuating pressure of the sound wave into an electric signal that corresponds in voltage to the pressure of the sound wave. An electric-to-acoustic transducer in a telephone receiver reconverts the signal to sound.

Defects in Telephone Transmission

The measure of signal power is the deciBel, more commonly abbreviated dB and pronounced *dee bee*. This is a logarithmic unit

Signal Power Compared to DeciBels

RATIO OF SIGNAL POWERS	POWER (dB)
100,000,000,000:1	110
1,000,000:1	60
10,000:1	40
100:1	20
1:1	0
1/100:1	−20

named for Bell. If any two powers P_2 and P_1 are being compared, the power ratio measured in deciBels is

$$10 \log_{10}\left(\frac{P_2}{P_1}\right) \text{dB}$$

One deciBel corresponds to a power ratio of approximately 5/4; 3 dB is a ratio of about 2/1, and 10 dB is a ratio of 10/1. The range of powers between a barely audible sound and a painfully loud sound is about 110 dB.

Imperfect data indicate that the first telephones produced a sound at the receiver that was about 40 dB weaker (smaller in power in the ratio 1/10,000) than the speech into the transmitter, even when the transmitter was connected directly to the receiver. At a distance of some miles the received signal was far weaker still. Only users who spoke loudly and clearly were understood.

But a reproduced sound differs from an original voice in ways other than power and loudness. One of these differences had to do with frequency. We shall see in more detail in Chapter 3 that complex sounds, voice or music, are made up of components of low frequency (low pitch) and high frequency (high pitch). If the low frequencies are not transmitted, the reproduced sound is tinny, like the sound of a pocket transistor radio. If the high frequencies are not transmitted, the reproduced sound is dull and muffled and hard to understand. If too narrow a range of frequencies is transmitted, it is difficult or impossible to understand the reproduced speech.

The response of the human ear to sound varies with the frequency and the power or intensity of the sound. In the diagram on this page, each curve represents sound of equal loudness. The ear is most sensitive to sounds with a frequency of about 3500 Hz. Sounds with frequencies below or above 3500 Hz are more difficult to hear unless they have a greater intensity. Thus, to the human ear a signal with a frequency of 3500 Hz and a sound intensity of 2 dB is equal in loudness to a signal with a frequency of 100 Hz and a sound intensity of 44 dB; the ear is not as sensitive at 100 Hz as it is at 3500 Hz.

Bell's first transmitters and receivers emphasized some frequencies and failed to transmit others. It was difficult to understand the re-created speech not only because it was weak, but also because important frequencies were missing.

A third factor that affects the quality of telephone transmission is noise. In long-distance calls you may hear a hissing or shushing sound even when no one is speaking. The sound is caused by random electric signals that are unavoidably added to the electric speech signal when it is sent over telephone circuits. Always, *some* noise is there, whether or not it is strong enough to be heard, or noticed.

If P_S is the signal power and P_N is noise power, the signal-to-noise ratio expressed in deciBels is

$$10 \log_{10} \left(\frac{P_S}{P_N} \right) \text{dB}$$

A speech signal-to-noise ratio of 60 dB or more is really high fidelity. A signal-to-noise ratio of 20 dB is very intelligible but noticeably noisy. Between 10 and 1 dB the signal becomes nearly unintelligible.

Today's long-distance transmission is increasingly by digital signals over optical fibers, and noise does not accumulate with distance. The signal-to-noise ratio is "designed in," so to speak, and is determined by the number of binary digits used per sample (we shall discuss digital transmission later).

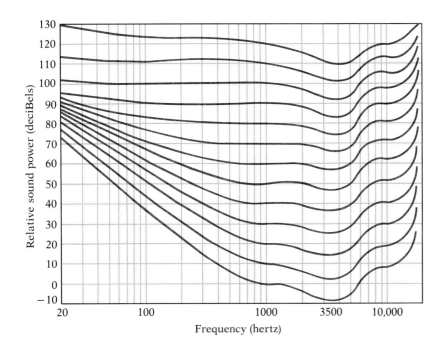

Our sensitivity to sound depends on the frequency of the sound. Each curve represents sounds of equal loudness. The top curve represents the threshold of pain, about the intensity at a loud rock concert at its loudest moments. The bottom curve represents the threshold of hearing. At 3500 Hz the separation of the threshold of hearing from the threshold of pain is about 110 dB, a ratio of 10 to the eleventh power. CAUTION—the curves shown are for healthy young people who have not had their hearing damaged by intense sounds.

Harvey Fletcher (right), a giant of psychoacoustics and onetime president of the American Physical Society, and J. C. Steinberg (left), noted researcher in the field of speech quality, demonstrate an electronic talking apparatus that could produce vowel sounds.

Standards of Speech Quality

Satisfactory transmission of the human voice by means of the telephone was not easy to achieve. It depended on making excellent transducers to transmit and receive signals and amplifiers to strengthen signals. Equally important, engineers had to know what quality signal would be easily intelligible to listeners. Transducers and transmission had to conform to standards based on studies of the intelligibility, or articulation, of telephone signals.

When I first went to Bell Laboratories, I visited a laboratory where women spoke strange sentences into a telephone, and listeners wrote down what they heard. "What did thed do?" If the volume was too weak or if the range of frequencies transmitted was too narrow, the listener might write down "sed" rather than thed."

The experimenters were making articulation tests to determine the accuracy of sound transmission required in telephony. They were working under the direction of Harvey Fletcher, a great and widely known student of speech and hearing. (A reference to "George" Fletcher appears in the English translation of Aleksandr Solzhenitsyn's *The First Circle*. One of the characters says to another concerning the latter's work on articulation, "I am terribly sorry that your original monograph was printed in a small classified edition, depriving you of the glory of being recognized as a Russian George Fletcher." Solzhenitsyn or his translator had converted Harvey to George.)

The sophisticated work of Fletcher and his colleagues was done to set reasonable standards for the quality of telephone transmission. It is costly to transmit a very broad band of frequencies or to reduce to a very low level the noise added in transmission. A signal that is too weak is difficult to understand; a signal that is too strong is annoying. How loud and how faithful should a signal be in order to be quite satisfactory without being too costly? What you hear on the telephone is an attempted answer. It sounds strange compared with high-fidelity radio, but it is quite intelligible.

Fletcher was able to perform his crucial experiments only because refined apparatus had been developed that could accurately convert acoustic signals into electric signals, and accurately convert amplified electric signals back into acoustic signals.

It was not easy to make transducers that would respond accurately over a wide range of frequencies. In telegraphy the telegraph key is a transducer because it turns an electric current on or off in response to the motion of the hand. The telegraph key need operate only at the speed of the hand, and it need merely turn an electric current on and off. The transducer at the receiving end is the telegraph sounder, which clicks when the current is turned on or off. A transducer of sound waves has a far more difficult task. A sound wave of speech fluctuates rapidly and smoothly with time, and so must the electric current that represents it in telephony. Indeed, this undulatory character of both sound and the telephone current was a chief claim of the patent that Bell filed on February 14, 1876. Bell cited the undulatory nature of the current in his telephone as clearly distinguishing telephony from telegraphy.

Transmitters and Receivers

In telephony the transducer at the transmitting end must produce a rapidly varying undulatory current, not an on-off current, and the transducer at the receiving end must respond to rapid variations in current and produce a smoothly varying sound pressure. In his early telephones Bell used the same device at both transmitting and receiving ends. The device consisted of a permanent magnet, a coil, and an iron diaphragm.

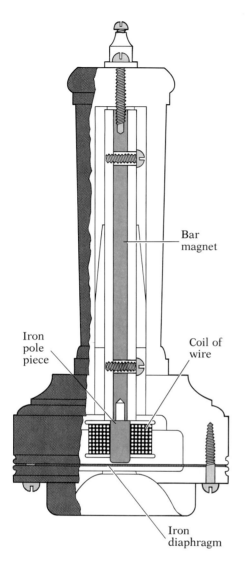

Bar magnet

Iron pole piece

Coil of wire

Iron diaphragm

In the first telephones the same device was used as a receiver and as a transmitter. The receiver-transmitter (shown in cutaway section) consisted of a bar magnet with an iron pole piece at the end, a coil of wire wrapped around the pole piece, and a thin iron diaphragm close to the end of the pole piece. When the device operated as a receiver, the current flow in the wire increased or decreased the pull of the magnet on the diaphragm, causing it to vibrate and emit sound. When the device operated as a transmitter, sound caused the diaphragm to vibrate, changing the magnetic flux through the coil, and producing a voltage between the ends of the wire of the coil. The receivers of the first telephones I saw and talked over looked much like this first receiver.

Some transducers, like the telegraph key, merely switch an electric current on and off. But others actually convert acoustic power into electric power, as an electric generator converts mechanical power into electric power.

Bell's telephone receiver and transmitter was this kind of device. It made use of two related principles of electricity and magnetism. One is that a change in a magnetic field through a coil of wire produces a voltage between the ends of the wire. The other is that a current in a coil can increase or decrease the strength of a magnetic field and hence the attraction of iron by the magnet.

In Bell's device a coil of wire was wrapped around an iron pole piece at the end of a bar magnet. Close to the end of the pole piece was a thin iron diaphragm. When the device operated as a transmitter, sound caused the diaphragm to vibrate. Motion of the iron diaphragm toward or away from the permanent magnet strengthened or weakened the magnetic field through the coil, and this change produced a voltage between the two ends of the wire. Hence, Bell's receiver turned an acoustic signal into a corresponding electric signal. When the device operated as a receiver, the fluctuating current flowing in the coil increased or decreased the strength of the magnetic field and its pull on the diaphragm, causing it to vibrate and emit sound.

Bell's device has continued in use as a receiver, although it has been changed somewhat throughout the years. Winding two coils around the pole pieces at the ends of a curved (horseshoe) magnet was found to be more effective than one coil around a pole piece at the end of a bar magnet. The exact configuration of the telephone receiver changes, but the principle of today's telephone receiver is the same as Bell's. Bell's receiver, used as a transmitter, or microphone, is the progenitor of various forms of dynamic, or magnetic,

CHAPTER 2

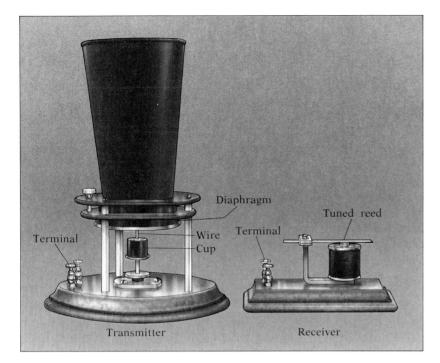

Terminal

Diaphragm

Wire

Cup

Terminal

Tuned reed

Transmitter

Receiver

In Bell's liquid transmitter, a speaker's voice caused a diaphragm to vibrate; a small wire attached to the center of the diaphragm moved up and down in a metal cup of acidulated water. As the wire moved farther in or out of the water, the electric resistance between the wire and the cup changed. When the terminals of the transmitter and the receiver were connected in series with a battery, the changes in resistance caused a changing current to flow through the receiver. The motions of the transmitter diaphragm were reproduced as motions of a tuned reed in the receiver. The liquid transmitter was a good idea—sound power used to control electric power rather than to produce it—but in practice the changes in resistance were not large enough.

microphones that have been used in sound systems, including high-fidelity systems.

Compared with a telegraph key, Bell's device had a great weakness as a transmitter. The motion of the key can be used to turn a large current on or off. The vibrations of Bell's iron diaphragm could only turn a feeble acoustic signal into a still-feebler electric signal. Bell was aware of the problem and tried to overcome it. In the spring of 1876 he built a "liquid" transmitter to control an electric current produced by a battery rather than generate an electric current all on its own. But he abandoned the liquid transmitter because the electric signal that it produced was too weak. Yet the general idea was sound. Bell's liquid transmitter operated on the same principle as the telephone transmitter of today, the carbon transmitter (or carbon microphone).

The carbon transmitter was a real breakthrough. The power of its electric signal can actually be greater than the acoustic power of the input voice signal. Thus, the carbon transmitter acts as an amplifier. The carbon transmitters in use today follow a patent filed in 1886 by Thomas A. Edison, a truly many-sided genius. A

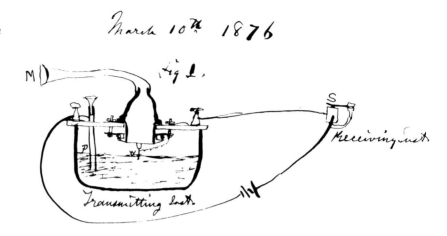

Bell's liquid transmitter, sketched here in Bell's original notes. Edison's carbon transmitter used a similar approach—and worked.

thin iron diaphragm in the transmitter moves in response to the pressure of a sound wave. That motion varies the pressure on carbon granules confined between two electrodes. An increase in pressure causes a decrease in the electric resistance of the granules; a decrease in pressure causes an increase in the electric resistance. Thus, if a constant voltage is applied between the two electrodes, the variation of the resistance in response to sound pressure causes the electric current flowing from one electrode to the other to vary in almost exact accord with the sound pressure at the diaphragm.

The carbon microphone is used because it has become very cheap to make and because it produces a strong signal. This last feature was especially important in the early days of telephony, before the vacuum tube, and later the transistor, made it possible to amplify electric signals. The defects of the carbon microphone are that it is noisy (it adds a hiss to the desired signal), it produces distortion (the electric signal does not correspond faithfully to the acoustic signal), and it responds somewhat unequally to different frequencies.

Advances in Sound Quality

In the early days of telephony, the chief problem was to make signals powerful enough to span greater and greater distances. The solution was to design transmitters that would produce a powerful

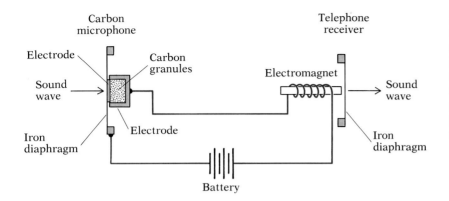

Carbon microphone

Electrode

Sound wave

Carbon granules

Electrode

Iron diaphragm

Battery

Telephone receiver

Electromagnet

Sound wave

Iron diaphragm

The carbon microphone, or carbon transmitter, is used in some telephone sets to this very day. The motion of an iron diaphragm varies the pressure on carbon granules confined between two electrodes. The variation in pressure causes the electric resistance between the electrodes to vary, which in turn varies the current.

electric signal and receivers that would produce the strongest possible sound from an electric signal. In early transmitters and receivers, quality was sacrificed to loudness. But for some purposes, high-fidelity sound may be needed. For example, to perform Fletcher's articulation tests one had to start with an almost perfect signal and degrade it in controlled ways.

How was one to get an almost perfect signal? The first requirement was a transmitter or microphone that would produce a truly accurate electric replica of an acoustic wave. By 1917 E. C. Wente had produced a truly high-fidelity microphone called a condenser microphone, which responded equally well to all frequencies from 20 to 8000 Hz. The condenser microphone makes use of a fundamental property of electricity that links electric field and electric charge. The electric field required in a condenser microphone must be produced by a rather high voltage, around 100 volts.

In a condenser microphone a thin metal foil is stretched parallel to a flat metal electrode or plate. These two elements or plates form what is called a *capacitor,* or what used to be called a condenser. A capacitor can store electric charge, a positive charge on one plate and an equal negative charge on the other plate. A capacitor has a capacitance, measured in farads and ordinarily designated by the letter C. The capacitance of the capacitor formed by two parallel plates is inversely proportional to the separation of the plates. If the plates are moved closer to one another the capacitance is decreased. There is a fundamental relationship between

In the condenser microphone, a vibrating foil causes the capacitance between foil and electrode to fluctuate. The fluctuating capacitance produces a fluctuating electric voltage that varies in accordance with a sound wave.

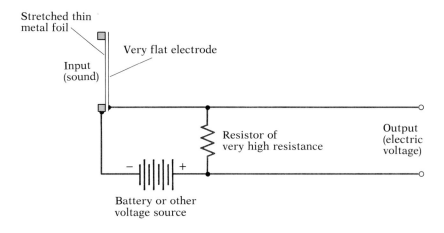

Stretched thin metal foil

Very flat electrode

Input (sound)

Resistor of very high resistance

Output (electric voltage)

Battery or other voltage source

the voltage V between the plates of a capacitor and the charge Q stored in the capacitor:

$$V = \frac{Q}{C}$$

In the condenser microphone a charge is put on the capacitor formed by the foil and the plate by connecting the foil to one pole of a battery or some other high-voltage source and the plate to the other pole. One connection to the voltage source is through a resistor of such a high resistance that the charge on the capacitor stays essentially constant over time. When a sound wave causes the foil to vibrate, the charge on the capacitor doesn't change but the capacitance does, and, by the relation among charge, voltage, and capacitance, the voltage must change in accord with the vibration of the stretched foil.

The second requirement for an almost perfect signal was an excellent receiver. Perhaps the best example is the moving-coil receiver, patented by Wente and A. L. Thuras in 1930. It had an almost flat frequency response from 10 to 8000 Hz—that is, it responded equally to frequencies over a broad range. The moving-coil receiver could produce a large acoustic output that faithfully reproduced the electric signal.

The moving-coil receiver is based on the principle that a current flowing in a conductor in a magnetic field produces a force on the conductor. In the moving-coil receiver, the conductor is a small

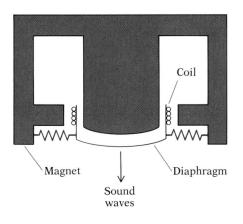

Coil

Magnet

Diaphragm

Sound waves

The moving-coil receiver sends a fluctuating current through a small coil in a magnetic field. The resulting force causes a diaphragm to vibrate, generating a sound wave.

CHAPTER 2

coil attached to a sensitive diaphragm. A fluctuating current representing a voice signal flows through the coil. The force produced causes the diaphragm to vibrate with the current, and so generates the sound wave the current represents. The high-fidelity cone speakers used in sound systems operate on this principle.

Although the telephone transmitters and receivers of today are the same in principle as those of Edison and Bell, they are far, far better in performance. Through the work of men like Wente they have been thoroughly understood, and well designed, and well tested.

As transistors and integrated circuits continue to reduce the cost of amplifying and processing speech, the carbon transmitter

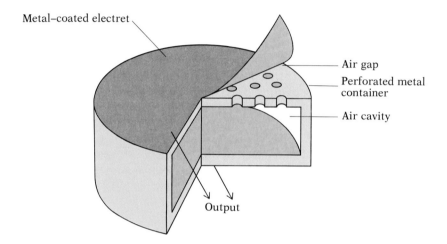

The electret microphone, like other condenser microphones, makes use of the pressure of a sound wave to produce vibrations in a thin, flexible diaphragm. The diaphragm is a thin sheet of plastic that has been allowed to solidify in a strong electric field, so that a permanent electric charge has been "frozen in" rather than supplied by a battery, as in the condenser microphone. A thin layer of metal is deposited on top of the sheet. The metallized dielectric sheet, or foil electret, rests lightly on the perforated surface of a cylindrical metal container. Because of surface irregularities, air gaps remain between the foil electret and the metal container. As incoming sound vibrations cause the foil electret to move, producing variations in voltage at the output. Perforations in the surround surface of the container allow communication between the large air cavity and the air gap, permitting the diaphragm to move but leaving it uninfluenced by sound pressure on the side that is shielded by the container.

will be completely replaced by less efficient but higher-fidelity devices. I would bet on the electret microphone, a simpler and less costly relative of the condenser microphone that is universal in home audio equipment.

Influences beyond Telephony

Bell System work on high-quality transducers had many offshoots. In 1925 workers at Bell Laboratories designed the orthophonic phonograph. It was the first mechanical phonograph to give an accurate reproduction of high-frequency and low-frequency sounds. Records for the orthophonic phonograph were made by picking up the sound electrically, amplifying the signal, and feeding it to an electrically driven recording stylus.

In the early 1920s radios and sound systems had sound quality comparable to that of today's pocket transistor radios. At that very time, through their interest in hearing and high-quality sound, Fletcher and his associates (including Wente) produced a multi-channel (stereo), high-fidelity sound system that transmitted by wire and reproduced at a distant location sounds with a frequency range, or bandwidth, of 15,000 Hz and a sound-intensity range of 80 dB. In 1933 the system was demonstrated in Constitution Hall in Washington, D.C. Leopold Stokowski turned over the baton of

Conductor Leopold Stokowski (left) and Harvey Fletcher (right) examine the first stereophonic sound equipment in 1933.

The Vitaphone motion picture sound system took top billing over John Barrymore in this 1926 billboard.

the Philadelphia Orchestra to an assistant and manned the controls of the sound system. The sound from the loudspeakers seemed just like the sound of the orchestra itself.

As early as 1922, Bell workers experimented with synchronizing a phonograph record with a silent film. The Vitaphone sound motion-picture system grew from that work. In the long run the Bell System found its role in radio, and later in television, to be that of transmitting programs from studio to transmitter, and from station to station, forming the radio and television networks that span the country. Through a combination of good sense and federal pressure, Bell divested itself of Electric Research Products, Incorporated, a subsidiary of AT&T, and got out of the talking picture business. But all through the days of the Bell System, and beyond, Bell Laboratories maintained its interest in high-quality sound and in the production of speech and its perception through the sense of hearing.

Telephony had the potential to become a universal service available directly to all because, in contrast to the telegraph or even the teletypewriter, anyone could use a telephone without training. Telephony has succeeded because from the start transmitters and receivers have been cheap and simple. The complexity of telephony lies in the complexity of interconnecting many phones, and this difficulty was surmounted through science and technology as the telephone network grew.

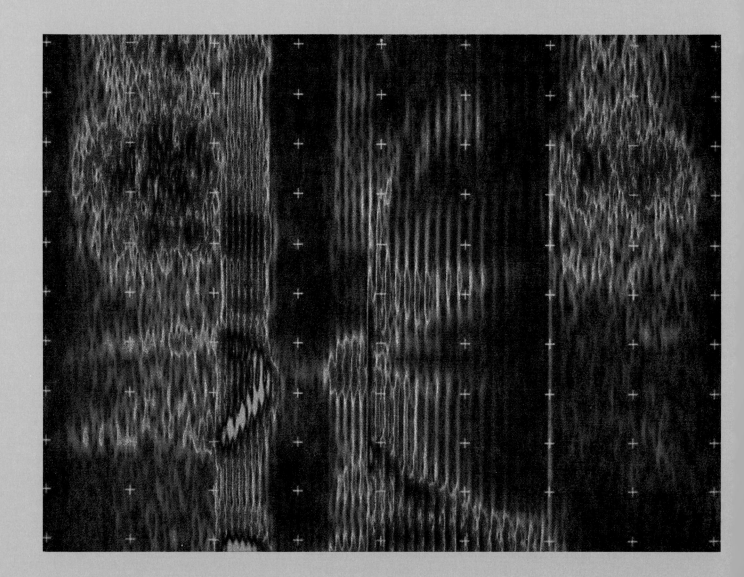

3
· · ● —

Theories of Communication:
Fourier to Shannon

Messages are sent from some source—a person or machine—to a receiver—also a person or machine. A message may be a display of text or diagrams, or it may re-create a human voice or a picture. Whatever its format, we want to reproduce the message in that format at the receiving end.

The signals that convey messages are all electric currents or voltages that vary with time. Signals are in a profound sense shaped like the messages they convey. The way the signal current or voltage varies with time may embody the message itself, although the message can be hidden by a complicated encoding. Complicated signals are difficult to deal with, both conceptually and mathematically. Yet it is possible to characterize a message in a way that will give us a deeper insight into what is communicated and how it is communicated.

There are two very different ways to characterize a message, born in very different eras. In the early nineteenth century Jean

A color spectrum of a voice pronouncing the word signals. *As a word is spoken into a sound spectrograph, its spectral composition is recorded and analysed. The machine plots time along the x-axis and frequency along the y-axis. Color corresponds to the intensity of the spectral energy.*

Baptiste Joseph Fourier, a French mathematician and physicist, discovered a wonderful thing while investigating the flow of heat. Any varying signal, no matter how complicated, can be represented by a set of simple waves, each with its own frequency. The waves are sine waves, curves that rise and fall periodically with time. The signal can be thought of as a sum of sine waves whose frequencies lie within a given range or band.

In 1948 Claude Elwood Shannon, the American mathematician, gave us his mathematical theory of communication, which we now call information theory. Shannon gave a precise mathematical meaning to information, and he gave methods of measuring the rate at which a message source generates information and the rate at which a communication channel can transmit information. We may regard Shannon's characterization of the process of communication as final. But Shannon's ideas summarize and clarify all that had gone before; we can neither appreciate nor apply his work without understanding the ideas that stemmed from the work of Fourier.

Fourier and Frequency

Fourier's discovery that any waveform can be accurately represented as a sum of sine waves was astonishing to his contemporaries. Subsequent workers in the fields of music, speech, and electricity have found his discovery indispensable.

A sine wave is a very smooth, simple curve. As illustrated in the diagram on the facing page, a sine wave can be described by the height of a crank on a wheel that rotates at a constant angular velocity. A plot of the height of the crank against time is a sine wave. In order to describe a sine wave completely we need to give just three quantities: amplitude, frequency, and phase. Amplitude is the value of the wave in either the positive or the negative direction. When describing the sine wave, we give the peak amplitude, the maximum value of the wave in either direction. But sometimes we may want to know the instantaneous amplitude, the amplitude at some particular instant. Frequency is the number of complete cycles per unit of time, specified in hertz (Hz), or cycles per second. Phase is given by the exact time or the relative time at which the wave changes from negative to positive. Usually, phase is specified either in degrees (there are 360 degrees in one complete cycle) or in radians (there are 2π radians in one complete cycle). The term *sine wave* is used to refer to either the trigonometric sine curve or the

In his studies of the flow of heat, French physicist Jean Baptiste Joseph Fourier (1768–1830) revolutionized many areas of science by showing that a sum of sine and cosine curves of different frequencies can accurately represent any curve plotted against time.

CHAPTER 3

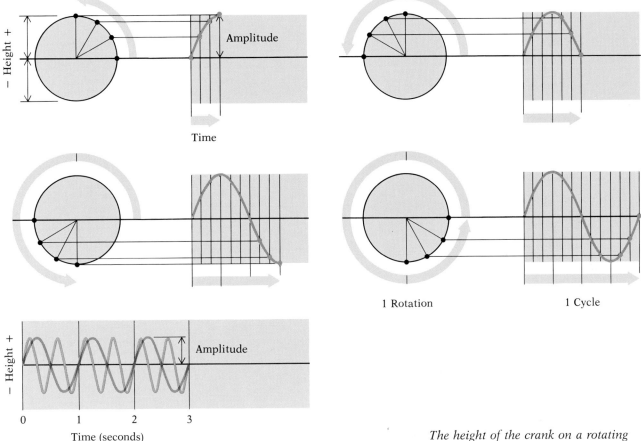

Amplitude

Time

1 Rotation

1 Cycle

Amplitude

0   1   2   3

Time (seconds)

trigonometric cosine curve. A sine wave with a phase such that its peak amplitude occurs at time zero is indeed a cosine curve.

The mathematics of communication signals and circuits provides a simple and very useful means for representing sine waves. A sine wave of a given frequency has a peak amplitude and a phase. The peak amplitude and the phase can both be represented by what is called a *complex number*. That is not strange, because a complex number is really a pair of numbers, ordinarily called the real and imaginary parts of the number. A complex number, however, can be treated in mathematical operations as a single number, and that is very important.

The height of the crank on a rotating wheel traces a smooth curve that describes a sine wave. The sine waves in the bottom panel of the illustration are identical in amplitude but differ in frequency. Both have the same phase in the sense that they change from negative to positive at time equal to zero.

Fourier Series and Transforms

There are two ways to represent voltages by means of sine waves. A Fourier series can be used to represent a voltage that repeats over and over, periodically and endlessly. A Fourier transform, or Fourier integral, can be used to represent a voltage that never repeats in time.

We can represent a function $f(t)$ of time, t, that extends from a time $t = -T/2$ to $t = T/2$ by the Fourier series

$$f(t) = \sum_{n=-\infty}^{n=\infty} C_n e^{j2\pi(n/T)t}$$

Here n is an integer such as -2, -1, 0, 1, 2, and so on, and j is $\sqrt{-1}$. We can find the coefficient C_n by means of the relation

$$C_n = \left(\frac{1}{T}\right) \int_{-T/2}^{T/2} f(t)e^{-j2\pi(n/T)t}\, dt$$

A Fourier series is always periodic, that is, it repeats itself every T seconds. If $f(t)$ is not periodic, the Fourier series representation above will not be correct for t smaller than $-T/2$ or larger than $T/2$. If $f(t)$ is periodic with a period T, the Fourier series representation will be valid for all time.

An electric signal must change unpredictably with time in order to convey a message. The representation of such a signal is achieved by a Fourier transform, or integral. If $v(t)$ is a function of time t, f is frequency and j is $\sqrt{-1}$ then the Fourier transform $V(f)$ of $v(t)$ is

$$V(f) = \int_{-\infty}^{\infty} v(t)e^{-j2\pi ft}\, dt$$

$V(f)$ specifies the phases and amplitudes of the sine waves that are the real part of $e^{j2\pi ft}$. The time function $v(t)$ can be recovered from the frequency spectrum $V(f)$ by the inverse Fourier transform

$$v(t) = \int_{-\infty}^{\infty} V(f)e^{j2\pi ft}\, df$$

Although $V(f)$, the Fourier transform of $v(t)$, is complex, the time function derived by this relation from the Fourier transform $V(f)$ of any real function will be real, in spite of the complex quantities in the integral.

It may seem odd to go from a known signal waveform to its representation in terms of sine waves. Why not take signal waveforms as they come, as they "really are"?

Our senses and many communication devices interpret signals by roughly analyzing them into component sine waves. The ear very clearly distinguishes between a sound containing sinusoidal components of high frequency only, waves that rise and fall frequently, and a sound containing components of low frequency only. A structure in the ear called the cochlea sorts out a sound wave into sinusoidal components of different frequencies, and to a degree we hear such components separately.

Hearing is not the only sense that responds differently to different sinusoidal components in a signal. Our sense of sight interprets light waves of different frequencies as different colors. The lowest-frequency light that we can see evokes the sensation of red,

Complex Numbers

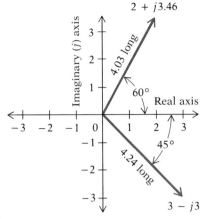

A complex number has the form $x + jy$, where x and y are real numbers and $j = \sqrt{-1}$. From this definition, $jj = -1$. We see that $1/j = j/jj = -j$. Complex numbers can be represented as points in the *complex plane*, where we plot the imaginary part as distance above a central point called the *origin*, which is located at $0 + j0$, and the real part as distance to the right of the origin. We can draw a vector, a line with an arrow at the end, between the origin and the point $x + jy$ which represents the complex number; this vector also represents the complex number. The length of the vector, the distance from the origin to the point, is $\sqrt{x^2 + y^2}$; this is the *magnitude* of the number. The angle between the vector and the real axis is the *phase* of the complex number. In the figure, two complex numbers are shown in this way. The first is $2 + j3.46$. The magnitude of this complex number is approximately 4.03, and its phase is 60 degrees counterclockwise from the real axis. This is called a leading phase. The other complex number is $3 - j3$. The magnitude of this complex number is approximately 4.24 and the phase is 45 degrees clockwise for the real axis. Complex numbers facilitate computations concerning complicated waveforms, such as communication signals. Both the amplitude and phase of a sinusoidal component of a given frequency can be represented by a single complex number. Complex numbers are also important in calculations concerning signals and circuits because the impedance of a circuit is a complex number which changes with frequency.

and light of increasingly higher frequencies evokes the sensations of orange, yellow, green, blue, and violet.

The ear and the eye can readily sort out and respond to those sinusoidal components of a complicated signal whose frequencies lie in one range, while rejecting those sinusoidal components of the signal whose frequencies lie in some other range. The same sorting out and rejecting, or filtering, occurs when we tune in the signal from one radio or television station and reject signals from other transmitters. Different stations send out signals made up of sinusoidal components that lie in different ranges of frequencies.

The frequency spectrum of a signal is obtained by plotting the peak amplitudes of the frequency components against their frequencies. As the diagram shows, in the simple case in which the signal is a sine wave, the frequency spectrum has a single amplitude at a single frequency. In contrast, in the frequency spectrum

The frequency spectrum of a signal depends on the frequencies and peak amplitudes of the sinusoidal components making up the signal. If the signal is a sine wave (A), the frequency spectrum of the signal is represented by a single frequency component at a frequency f_0 (and a single phase). If the signal is not sinusoidal but repeats periodically in time, the frequency spectrum consists of components at frequencies f_0, $2f_0$, $3f_0$, and so on at integer (or harmonic) multiples of f_0, each component with its own phase and peak amplitude. For the example periodic square wave (B), the frequency components are only at odd harmonics, f_0, $3f_0$, $5f_0$, and so on. If the signal is nonperiodic, that is, does not repeat in time (C), the frequency spectrum is continuous over the bandwidth of the signal and every possible frequency within the band is present.

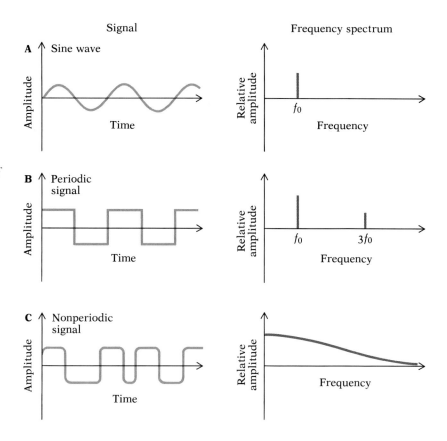

Signal   Frequency spectrum

of a complicated voice signal every possible frequency within the bandwidth is present. Another separate spectrum would be needed to show the phase of the frequency components of the signal, which is closely related to that of the baseband signal. The sprectral representation of a signal contains the information necessary to re-create the original time-varying signal.

Beyond all this, the representation of a complicated signal as a sum of sine waves makes both the understanding of communication and the calculations concerning communication simpler than they would otherwise be.

Sine Waves and Linear Circuits

Many traditional communication circuits are *linear* circuits, which are formed by a particular combination of resistors, which dissi-

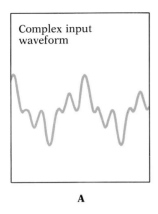

Complex input waveform

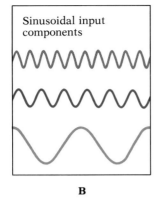

Sinusoidal input components

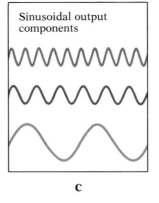

Sinusoidal output components

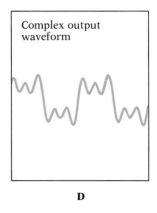

Complex output waveform

A   B   C   D

pate electric energy, capacitors, which store electric charge, and inductors, which store magnetic energy. The lines that carry communication signals are linear circuits, and so are good "linear" audio amplifiers. Linear circuits can also act as filters that sort out sine waves whose frequencies lie in a particular range or band. Today the functions of linear circuits are frequently performed by electronic devices, analog or digital.

The widespread presence of linear circuits explains in part why it is so practical to represent a signal as a sum of sine waves. That is the ideal way to describe the signal in a linear circuit, because if you put a sine wave into a linear circuit you get a sine wave out. For each sinusoidal component of a complex waveform that enters a linear circuit, a sinusoidal component of the same frequency comes out, although the circuit may change the relative phases and amplitudes of the sinusoidal components. Because the output amplitudes and phases may be different, the output waveform—that is, the way the output current of voltage varies with time—may look very different from the input waveform—the way the input current or voltage varies with time. For example, an electric current of short duration, a short pulse of current, may be lengthened in passing through a linear circuit. Because linear circuits transmit the frequencies put into them without producing new frequencies, signals formed of sine waves of different frequencies can be put through a linear circuit simultaneously without interfering with each other.

The idea of linearity can be illustrated very simply by a spring whose deflection is proportional to the weight stretching it. For

A linear circuit transmits any sinusoidal component of a complicated waveform as a sinusoidal output, but it may change the relative phases and amplitudes of the various sinusoidal components and thus change the shape of the complicated waveform that is the sum of all the sinusoidal components. For the linear circuit whose performance is illustrated in the figure, the complex input waveform A is the sum of the three sinusoidal components shown in B. This particular linear circuit transmits these frequency components without altering the amplitudes, but it does shift the phase of the component of lowest frequency, as shown in C. When the output components of C are added up they give the output waveform shown in D, which is clearly different from the input waveform A.

Linear Circuit, Ohm's Law, and Impedance

Ohm's law, $I = V/R$, tells us the current a voltage will cause to flow in a circuit—but only if the voltage and the current don't vary too rapidly with time. What if the voltage and current *do* vary rapidly with time? We can write a linear relation between voltage and current as

$$I = \frac{V}{Z}$$

Here Z is called the *impedance*. But, this relation applies only for a sine wave. Fortunately, we can represent any voltage waveform as a sum of sinusoidal components of different frequencies and apply the relation for each sinusoidal component separately. Then the total current will be the sum of the various computed sinusoidal currents.

The impedance Z is a complex number that varies with frequency. Let us consider some examples.

The impedance of a resistor of resistance R is simply R. R is a real number, and it is the same at all frequencies.

A capacitor is made up of sheets of foil separated by a thin dielectric, a nonconducting material. The impedance of a capacitor is $Z = 1/(j2\pi fC)$, where C is capaci-

tance in farads. For a sinusoidal voltage V across a capacitor, the sinusoidal current I is $I = j2\pi VfC$. Note that in complex notation the point representing I lies counterclockwise from the point representing V; the current *leads* the voltage in phase. And, the current increases with increasing frequency.

An inductor is a coil of wire. The instantaneous voltage across an inductor is equal to the rate of change of current through the inductor. The impedance of an inductor is $Z = j2\pi fL$, where L is inductance in henries. For a sinusoidal voltage V across the inductor, the sinusoidal current through the inductor is $I = V/(j2\pi fL) = -jV/(2\pi fL)$. The current decreases with increasing frequency, and it lags the voltage in phase.

Suppose that, as shown in the figure, we connect a resistor of resistance R, a capacitor of capacitance C, and an inductor of inductance L in series, and apply a voltage V of frequency f. The total impedance will be $Z = R - j/2\pi fC + j2\pi fL$, and the current I will be

$$I = \frac{V}{R + j2\pi fL - j/2\pi fC}$$

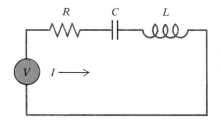

For a constant voltage V, let us consider how I varies with frequency. When $2\pi fL = 1/(2\pi fC)$, then $f = 1/(2\pi\sqrt{LC})$ and $I = V/R$. This is a resonant frequency; the current is larger at this frequency than at any other. When f is very small, the current is approximately $jV2\pi fC$. It leads the voltage in phase and decreases as frequency is decreased. For very high frequencies, the current is approximately $-jV/(2\pi fL)$; the current lags the voltage in phase, and decreases with increasing frequency. The circuit acts as a band-pass filter, transmitting most strongly frequency components whose frequencies lie near the resonant frequency. The band of frequencies passed becomes narrower as we decrease R.

such a linear spring, the amount of stretch produced by a five-pound weight is the sum of the stretches produced by a two-pound weight and a three-pound weight applied separately. Because the

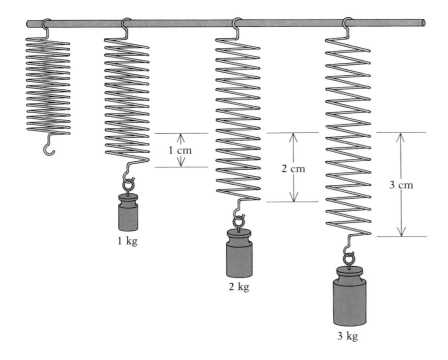

1 cm

2 cm

3 cm

1 kg

2 kg

3 kg

A linear relation is illustrated by a weighted spring. When a 1-kilogram weight is hung on the spring, the spring stretches 1 centimeter; when a 2-kilogram weight is hung on the spring, the spring stretches 2 centimeters, and so on. The distance the spring stretches is strictly proportional to the weight hung from it; thus, the relation between the amount the spring stretches and the weight hung on it is linear. Further, we should note the following consequence: the distance that a 3-kilogram weight stretches the string is the sum of the distance that a 1-kilogram weight stretches the string and the distance that a 2-kilogram weight stretches the string. For a linear system, the output (here, distance stretched) caused by the sum of two inputs (here, a 1-kilogram weight and a 2-kilogram weight) is the sum of the outputs (distances stretched) for two signals (weights) applied separately.

circuit is linear, the output amplitude is proportional to the input amplitude. If you double the input amplitude, you double the output amplitude.

Bandwidth

The Fourier representation of signals as a sum of sine waves strikes at a particularly vexing property of signals—the rapid and "unexpected" changes that a signal must undergo in representing a message.

An electric current can continue to rise and fall smoothly and periodically forever and in a sense never really change. It simply follows a repetitive pattern "mechanically." We can predict what it will be at any time in the future. A note whistled forever at a constant loudness and pitch is just a note; it cannot be a tune or convey a message. An electric signal must change unexpectedly with time in order to convey information that we do not already know.

**Comparison of Telephone Bandwidths and
Other Modes of Communication**

MODE OF COMMUNICATION	BANDWIDTH (Hz)	RATIO OF BANDWIDTH to telephone band
Telex	200	0.05
Telephone connection	4,000	1
High fidelity	16,000	4
Television	4,000,000	1,000

In representing a signal that can change both rapidly and unexpectedly we must use a collection of sinusoidal components whose frequencies span some frequency range, or bandwidth. The bandwidth of a signal is the range of frequencies that must be used to represent it. If frequency components ranging from 0 to 4000 Hz, for example, are needed to represent a signal, the bandwidth of the signal is 4000 Hz.

There is a fundamental relationship between how rapidly and unexpectedly a signal can change with time and the bandwidth of the signal. A slowly varying data signal can be represented adequately by a bandwidth of sinusoidal frequency components 200 Hz wide. A more rapidly varying telephone signal requires a larger bandwidth of frequencies, roughly, 4000 Hz. A high-fidelity audio signal requires a bandwidth of about 16,000 Hz. A still more rapidly varying television signal requires a greater bandwidth, roughly, 4,000,000 Hz.

Many sorts of signals can be sent over a telephone connection, including data and fax as well as voice. Thus, the bandwidth of the telephone connection, which is about 4000 Hz (really, 200 to 3200 Hz), is a convenient standard of comparison. Facsimile can be transmitted over a telephone connection at a rate of about one page a minute. Text can be transmitted about 10 times as fast.

The Sampling Theorem

We have used the representation of signals by bands of sine waves in seeking to understand how rapidly and unexpectedly a signal can change with time. A mathematical theorem called the sampling theorem tells us that a signal of bandwidth B Hz can have $2B$ independent and successive amplitudes each second. Or, conversely, if we insist that a signal waveform have successively $2B$ independent amplitudes in a second, it must have a bandwidth of B Hz.

An example from television transmission demonstrates the relationship between bandwidth and the number of amplitudes transmitted each second. The image on a television screen is composed of small dots of light called pixels, which are arranged in horizontal lines. The lines of pixels compose the frame, or one picture image. Variations in the brightness of the pixels determine the picture we see. Consider a television transmission with a bandwidth of 4,000,000 Hz that sends a complete 525-line frame 30 times each second. Ideally, using this bandwidth allows us to send 8,000,000 distinct amplitudes per second. The possible number of distinct brightnesses of pixels that can be sent per second is thus 8,000,000. If we divide 8,000,000 by 30 (frames per second) and again by 525 (lines per frame), we arrive at about 500 pixels per line. When TV standards were set, that number seemed to give both reasonable cost and quality of reproduction. Putting things the other way around, if we want to send 500 independent pixel brightnesses per line, we must use a bandwidth of 4,000,000 Hz or more.

In actual practice it is usually too complicated and costly to use a signal of bandwidth B to send as many as $2B$ distinct amplitudes. Components of reasonable cost and complexity, and practical means of encoding or modulation, cut the number down somewhat (encoding and modulation are discussed in Chapter 4). Thus, the number of message-conveying amplitudes transmitted per second may fall considerably below twice the bandwidth measured in hertz.

Noise

We attain a truly profound insight by thinking of signal waveforms as made up of sinusoidal components spanning some range, or band, of frequencies. It enables us to classify signals accurately as varying more or less rapidly and unexpectedly, and as requiring more or less bandwidth for transmission. But bandwidth is not the only coin of the communication realm. There is also power. The amount of power it takes to send a signal so that it is received

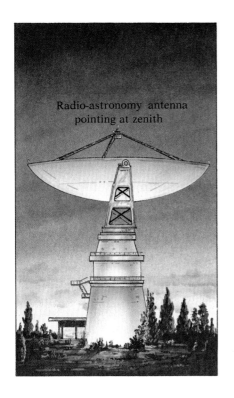

Radio-astronomy antenna pointing at zenith

Transmitter

Microwave antenna pointing at a wooded slope

Thermal, or Johnson, noise is proportional to the temperature of the noise source. A radio-astronomy antenna pointed at the zenith receives noise from interstellar space, which has a temperature of about 3 K ($-273°C$). During a cloudburst the antenna receives added noise from the comparatively hot rain. A microwave communication receiver pointed at a transmitter on a wooded hillside receives noise from the trees, which have a temperature of about 293 K ($20°C$).

clearly is related to bandwidth and to another factor, which is noise. You will recall from Chapter 2 that noise signals are unwanted signals that are always present in a transmission system. It is noise that makes a received signal fuzzy and uncertain.

There are various sources of electric noise. Thunderstorms cause a radio receiver to pop, and they create flashes on a television screen. A car radio becomes noisy when we drive under a high-voltage transmission line. Automobile ignition, fluorescent lamps, and other sources produce a different-sounding noise, shorter and sharper. The transistor amplifiers and other components in transmission systems add noise to the signal.

Some noise, such as that produced by automobile ignition, is avoidable. Other noise, such as that produced by transistor amplifiers in radio or television receivers, can be reduced. From year to year the all-frequency noise, or *white noise*, that is added to signals becomes less as we learn to make better and better amplifiers.

One sort of noise is completely unavoidable because, like light and heat, it is produced by all objects whose temperature is above

CHAPTER 3

Receiver

absolute zero (−273.15°C). It is called thermal noise, or Johnson
noise, after J. B. Johnson of Bell Laboratories, who discovered it in
1928. If we turn the antenna of a sensitive radio-astronomy receiver from the zenith down toward the horizon, the noise in the
output will increase many fold because of noise emitted from the
hot earth. If we point the antenna of such a receiver at the sun,
which is very hot, the noise output will go up by a factor of about
200. If we point the antenna anywhere in space, we will receive
noise that corresponds to 3 K (or kelvins; that is, degrees Celsius
above absolute zero). The radiation causing this noise is a true
vestige of creation, left over, we are told, from the big bang with
which the universe began some 20 billion years ago. In 1978 Arno
A. Penzias and Robert W. Wilson of Bell Laboratories were
awarded the Nobel prize for the discovery of this cosmic microwave background radiation.

   Because Johnson noise is proportional to receiver bandwidth
as well as to source temperature, more total noise power is added
to the signal in receiving broadband signals than in receiving nar-

rowband signals. The bandwidth used in telephony is around 4000 Hz; that used in television is around 4,000,000 Hz, a thousand times as great. Thus, the noise that is added to a television signal during transmission and reception is about a thousand times as great as that added to a telephone signal.

When we compare the unavoidable noise in various sorts of communication, it is the ratio of the total signal power to the total noise power that counts. We must use more power in transmitting a broadband signal than in transmitting a narrowband signal if we are to have a satisfactory ratio of signal power to noise power. In fact, if we wish to hold the signal-to-noise ratio constant, the transmitter power must be made proportional to the signal bandwidth.

We can tolerate different ratios of noise power to signal power in different sorts of communication. In transmitting on-off telegraph or data signals, a signal power that is 15 dB (32 times) greater than the noise power will ensure that the received code will almost never be misinterpreted. A voice circuit with a signal-to-noise ratio of 20 dB is intelligible but noisy; a 40-dB signal-to-noise ratio is good. High-fidelity recordings can have a signal-to-noise ratio of around 60 dB. Compact disc recording gives a signal-to-noise ratio of around 90 dB. It would take a yet better signal-to-noise ratio to match the capabilities of the human ear, which can hear over a power range of as much as 120 dB.

Shannon's Information Theory

An understanding of signals as sums of sine waves, and of bandwidth, noise, and signal-to-noise ratio, is essential to an understanding of the whole process of communication, especially of the theory that treats it mathematically.

Shannon's communication theory, or information theory, received the accolade accorded to many scientific discoveries—a wide application and misapplication in large areas of life and experience. Evolution gave us the evolution of society, of morals, and of science, as well as the origin of species. Relativity gave us the relativity of art, social systems, and morals, as well as the reconciliation of Maxwell's equations with the laws of motion. Information theory has been invoked in psychology, pedagogy, art, and who remembers what else?

Make no mistake. Information theory is not nonsense just because much nonsense has been written about it. Information theory has real content and value in the field of communication. It does not tell us everything we need to know, but it does tell us about ultimate limits that are true, useful, and surprising.

In order to treat the process of communication mathematically, Shannon did not merely examine different sorts of signals, their frequencies, their bandwidths, and the noise added in transmission. He did something bold and unprecedented.

He went behind the communication system itself. Shannon created a measure of the rate at which information is generated by a source that produces messages we wish to transmit. He gave a measure of the information-transmitting capacity of a communication channel such as a microwave system or a coaxial cable with repeaters (amplifiers) or an optical fiber system. He proved a fundamental theorem that states that if the source rate is less than or equal to the channel capacity, messages from the source can, in principle, be transmitted over the channel essentially without error (with less than any assigned error rate, however small). Shannon showed that error-correcting codes exist. By using such codes it is possible to attain virtually error-free transmission over noisy communication channels.

Examples will give us a feel for Shannon's theory. A message source may produce text, sound, or the waving of hands. We wish somehow to re-create messages from such a source at a distance. Shannon's characterization of the information rate of a message source does not deal with the meaning or content of the messages that the source produces. Rather, his information rate is a measure of how difficult it is to transmit without error, or with an assigned fidelity, messages generated by the source.

What is a valid measure of information in this sense? It must be uncertainty as to what message the source will produce. If we know what a person will say before he says it, what he says can convey no information to us. The rate at which a message source produces information must be a measure of the unpredictability of the messages that the source produces.

Shannon's information theory is necessarily of a probabilistic, or statistical, nature. His theory does not tell us about the generation of a particular message and its transmission over a communication channel at a particular time. Rather, it tells us about the average behavior of a message source and of a communication channel.

Shannon includes the following elements in his model of the process of communication: an information source such as a human speaker or a scene at which a television camera is pointed; a trans-mitter, which encodes messages from the information source for

Here Claude Elwood Shannon is shown with a maze-solving mouse constructed in 1952 to demonstrate how switching relays operate in dial telephones. This device reflects Shannon's master's thesis on the application of Boolean algebra to relay circuits (Chapter 8). Shannon is a hero to all communicators.

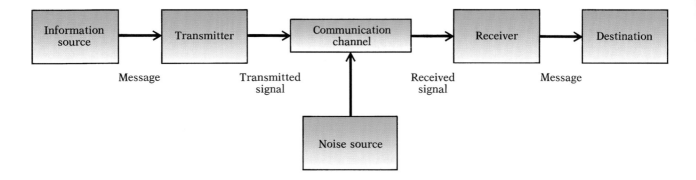

Shannon's diagram of the communication process.

effective transmission; a communication channel; noise, which modifies the encoded message unpredictably between transmission and reception; a receiver, which reconstructs the message from the information source with perfect or adequate accuracy; and a message destination, usually a listener or a viewer.

Entropy

Shannon's theory first establishes a quantitative measure of the amount of information in a message from a source. His analysis begins by characterizing the message source through a statistical analysis, that is, by measuring its inherent probabilities or predictability. By such analysis it can be shown of English text that *u* almost always follows *q*, that *e* is more frequently used than *z*, that the letter immediately following a period (the beginning of a sentence) is capitalized, and that *the* is more common than *elf*. But even though rules of predictability apply to some aspects of a text, the text as a whole is not entirely predictable. The degree to which messages from a source are unpredictable is taken as a measure of the amount of information in such messages. This measure is called the *entropy* of the source. The entropy is characteristic of the source that produces a message, and not of an individual message.

If we were gambling by phone, we might consider the outcomes of successive tosses of a coin as messages from a source (the process of tossing the coin). When we toss an honest coin, heads or tails comes up with equal probability. And the outcome of a particular toss is not influenced by the outcome of previous tosses. Shannon took the uncertainty of outcome of two equally likely possibilities as his fundamental measure of entropy, or amount of information. This unit of information has come to be called the *bit* (binary digit).

When we regard the sequence of heads and tails produced by successive tosses of a coin as messages from an information source, the rate at which the source generates information is one bit per toss. The outcomes of the tosses can be encoded as a sequence of the binary digits 0 and 1 by letting 0 represent heads and 1 represent tails.

What if the coin always turned up heads? Regarded as an information source, the coin would produce no information per toss, because the outcome would not be unpredictable, but would be known in advance.

The throws of an honest die (regarded as an information source) produce more information than the tosses of a coin, because any of six numbers can come up. In Shannon's terms, the information, or entropy, per throw of a die is $\log_2 6$, or 2.58 bits per throw.

The amount of information that a source generates depends both on how many different messages or outcomes the source can produce, and on how uncertain we are concerning which message the source will actually produce. If the source can produce only one message, it generates zero information per message. If it produces any of n messages with equal probabilities, the amount of information is $\log_2 n$ bits per message. If some messages are more likely than others, the entropy per message will be less than $\log_2 n$. Such message sources are said to be *redundant*.

How do we take redundancy into account in estimating the entropy, or uncertainty, of a human being writing text, for instance? Clearly, the writer does not write down just any letter next; that would produce gibberish. Shannon himself made experiments that showed that a person can guess with a high chance of success what letter will come next if given the text that precedes the letter. But although the chance of success in guessing the next letter is high, it is not perfect, because what a writer will write next is not entirely predictable.

Thus, we view the composer of music or the writer or the speaker as constrained by grammar, custom, and theme, but continually making unpredictable (or random) choices. It is the amount of entropy of this randomness, this unpredictability, that we must measure in order to measure the rate at which a redundant source produces information.

Information Content of a Text

Taking Shannon's model as correct, we can try to find the rate at which various redundant sources produce information. When data is transmitted between computers, letters, numbers, and other symbols are transformed into a special code, usually the American Standard Code for Information Interchange (ASCII). This code represents symbols by using 8 binary digits per character. We can transmit any English text by using the ASCII code, so that the number of bits per character actually needed—that is, the information in bits per character—must be less than eight. Indeed, if all 26 letters and the space occurred equally frequently, and if we were satisfied with capital letters and no punctuation, the rules we have already given say that we can transmit the text with only $\log_2$ 27 bits, that is, 4.8 bits per character, so that the information in bits per character must be less than 4.8.

Some letters occur in text more frequently than others. Taking this into account, it can be shown that the information rate must be less than 4.14 bits per letter. If we base our estimates on the frequency of occurrence of words rather than letters, we obtain an estimate of 9.7 bits per word, or about 2 bits per letter.

Shannon measured the uncertainty of the next letter in text by finding how well a person could guess it after looking at a certain portion of the preceding text. He estimated that the entropy of English text is between 1 and 2 bits per letter.

We can estimate the entropy of various message sources in a number of ways. When the signal is a waveform, we can measure the frequency of occurrence of signal amplitudes and of sequences of signal amplitudes and then measure the degree of success in predicting the next signal amplitude when past signal amplitudes are known.

According to Shannon's model, which agrees usefully with reality, a message source produces messages that have a certain entropy rate. The entropy can be measured in bits per message, bits per character, or bits per second. According to Shannon, messages from a source whose rate is n binary digits per second can literally be encoded into n binary digits per second. That is true only on the average. It is true for encoding a large succession of messages from the source.

Continuous Messages

Sources such as text that convey messages in discrete pieces always produce information at a finite rate. What about message sources, such as speech, that produce a smooth, continuous signal? Ideally, the signal voltage can have one of an infinite number of voltages at any moment. What, then, is to prevent the entropy of a continuous source from being infinite?

Continuous message sources, such as voice or pictures, have a finite rate only if we impose a *fidelity criterion*. A simple fidelity criterion can be fineness of detail, or signal-to-noise ratio. We simply count all voltages or brightnesses lying in some small range as the same; we do not try to reproduce voltage or brightness exactly in reconstructing the message for the recipient. A more sensible but intractable fidelity criterion is whether a whole speech or picture is acceptable to the destination. Such a criterion is almost impossible to evaluate.

Error Correction

Suppose that we do in fact or in principle reduce messages from a message source to some number of bits per second. We are faced with transmitting that number of bits per second over a communication channel. If the channel is a *discrete, noiseless* channel that transmits a finite set of characters without error, the problem is comparatively simple. If the channel can transmit without error any of N characters, the channel capacity is $\log_2 N$ bits per character. If the entropy rate of an information source is less than the capacity of a channel, then messages from the source can be transmitted over the channel without error. If the channel capacity is less than the entropy of the source, messages from the source cannot be transmitted over the channel without error.

I have pointed out that communication channels, discrete or continuous, are noisy. The signal received differs in an unpredictable way from the signal transmitted, usually because noise has been added to the signal during transmission. Shannon showed that even a noisy channel has a capacity measured in bits per second (or per character). He proved that if the information rate of a message source is less than the capacity of a noisy channel, messages from the source can be transmitted over the channel without error (strictly, with fewer errors than any assigned number, however small).

Channel Capacity

In his mathematical theory of communication, Shannon showed that even a noisy channel has a channel capacity measured in bits per second. Consider a communication channel of bandwidth B Hz. Suppose that the power received over the channel is a signal power P and a noise power N from noise somehow added during transmission and amplification. The channel capacity C in bits per second is

$$C = B \log_2\left(1 + \frac{S}{N}\right)$$

The equation shows the advantage of using a broad bandwidth to transmit messages.

Because noise is proportional to bandwidth, the thermal, or Johnson, noise power and bandwidth are related by the equation

$$N = kTB$$

where k is Boltzmann's constant and T is noise temperature in kelvins.

If N is all the noise that is added to the signal,

$$C = B \log\left(1 + \frac{S}{kTB}\right) \text{ bits per}$$
second

Note that in this equation the bandwidth B appears twice. For a given signal power, as B increases the channel capacity increases. Thus, if possible it is better to use more rather than less bandwidth in transmitting a signal with a given power.

As the bandwidth B is made larger and larger, the channel capacity approaches a limiting value

$$C = \frac{BS}{kT} \ln 2 \text{ bits per second}$$

The equation shows that the energy necessary to transmit one bit of information can never be less than $kT \ln 2 = 0.693 \, kT$ joule.

That limit is approached most closely in painstakingly designed communication systems for inter-planetary missions. To gain an idea of the energy involved, consider that the electricity used in lighting homes is measured by the kilowatt-hour. A kilowatt-hour is equal to 3.6×10^6 joules. Boltzmann's constant is equal to 1.38×10^{-23} joule per degree. Room temperature is about 293 K. At this temperature, $kT \ln 2 = 1.23 \times 10^{-27}$ kilowatt-hour per bit (or a decimal point followed by 26 zeros followed by 123). In principle, it does not take much power to transmit a bit of information. In practice it takes somewhat more—a few times more in well-designed systems.

The foregoing calculations are "classical"; they were derived without taking into account the quantum nature of electromagnetic radiation. I was surprised to find, some years ago, that $kT \ln 2$ joules per bit *is* the correct quantum-mechanical result, but it is almost impossible to approach closely in signaling with light. Still, workers at Caltech's Jet Propulsion Laboratory have attained signaling rates of several bits per photon.

To attain error-free transmission one must encode the message into a signal suited to the nature of the channel. Error-correcting codes have been devised for that purpose. For example, suppose a channel is a noisy binary channel, which transmits a sequence of binary digits (0, 1) but sometimes makes errors in doing so. A 0 may be received as a 1, or a 1 as a 0. If we wish to transmit n binary digits without error, we must send a signal consisting of more than n binary digits. In essence, we add error-check digits to the message digits to make up the signal that is to be transmitted. At the receiver, the correct message digits can be recovered even if some message and check digits, or both, are received in error. A crude everyday example of error correction—or at least, of error detection—is saying words twice to make oneself understood over a noisy channel, such as citizen's-band radio. This seems in accord with Jesus's injunction (Matthew 5:33–7) that in conveying a message one should not swear, but say Yea, yea; Nay, nay.

A simple error-correcting code is illustrated in the margin. This code makes use of a total of 24 digits to transmit 16 message digits. To show how the code works, the 16 message digits have been written in a 4-by-4 square. The check digits (shown in circles) have been written above the columns and to the left of the rows of message digits, so that there are overall five columns and five rows. The check digits are chosen so that the sum of the check and message digits in each overall row and each overall column is even. If one message digit is received in error, the sum of the digits in some row and in some column will be odd, and this locates the erroneous message digit, which can be changed. If one check digit is received in error, there will be an odd sum of digits in one row *or* column; this indicates an erroneous check digit rather than an erroneous message digit. The code shown is somewhat inefficient and would not be used in practice. Many error-correcting codes can correct more than one transmission error. Error-correcting codes are used in computers and in compact discs, as well as in digital transmission systems.

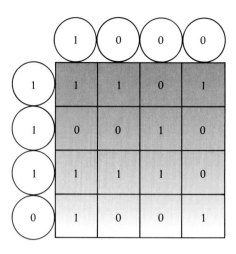

A simple error-correcting code makes use of a total of 24 digits to transmit 16 message digits. If one message digit is received in error, the sum of digits in some row and in some column will be odd, and this locates the erroneous message digit, which can be changed.

Shannon's Overall Process

Let us look at the overall process that Shannon proposed. Message sources are redundant. What comes from them seems more complicated or unexpected than it really is. Ideally, we should seek by a complicated encoding to represent the messages from the source by the smallest number of binary digits possible. If we succeed, the encoding of the source will produce a nonredundant stream of binary digits. We are then faced with the problem of transmitting a nonredundant stream of binary digits without error over a noisy channel. To do this we encode the nonredundant message digits into a redundant stream of signal digits. In sum, we remove redundancy from the messages produced by the message source and then add the right sort of redundancy to produce a signal that will give us error-free transmission over a particular noisy channel.

This Voyager *photograph of Neptune was transmitted to Earth as efficiently encoded digital data from a distance of some 4.5 million kilometers (2.8 million miles).*

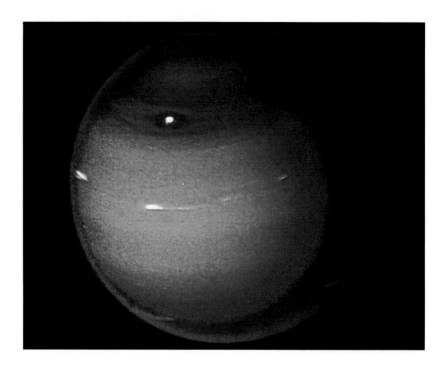

It is possible to come close to the ideal of an error-free transmission when it is worthwhile. Through error correction good pictures and reliable data come back from such spacecraft as *Voyager.* In a more mundane example, the almost complete absence of clicks or pops in playing compact disc recordings (which are digital) is attained through an error-correcting code. The code corrects for rare fabrication or reading errors that turn a 0 into a 1 or a 1 into a 0.

The Vocoder and Speech Encoding

There is another aspect of communication, well understood by Shannon and completely consistent with his theory, but not susceptible to mathematical attack. Sometimes there is a simple source behind a seemingly complex message. That is true of speech, for example. Our vocal cords produce sounds that are shaped by movements of our tongue, jaw, lips, and velum. When we speak, our vocal cords vibrate rapidly but fairly regularly, but our tongue, jaw, lips, and velum move rather slowly. We suspect that it should take less channel capacity to transmit the slow motions that shape the complicated speech wave than it does to transmit the speech waveform itself. And that is indeed so.

In 1936, long before Shannon's work, Homer Dudley of Bell Laboratories invented a device called the *vocoder*, which transmits intelligible speech without transmitting the speech waveform. Rather, the sending terminal analyzes the speech and sends signals telling whether the sound is voiced or unvoiced and describing the pitch of the sound and how the sound energy is distributed in frequency. At the receiving end, the slowly varying description of speech operates the controls of a speaking machine that produces synthetic speech in imitation of the talker at the transmitting end. The signals travel over a channel with less bandwidth than that of the original speech waveform.

This channel vocoder is not a good way to transmit speech efficiently, especially in a world of digital chips. Today, the channel vocoder has been superseded in speech transmission, and, indeed, in the generation of artificial speech for toys (Speak and Spell™), devices for the blind, and speaking cars and refrigerators, by a technique called *linear prediction*. In linear prediction, the next sample amplitude of the speech waveform is computed as a weighted sum of a number of preceding sample amplitudes. If the prediction is good enough, we don't have to transmit the sample amplitude. If there is a serious error in the prediction of the sample amplitude, a "correction" of some sort must be sent. I won't try to explain the intricacies of linear prediction. I will note that linear prediction is related in a fundamental way to filtering the speech waveform, in a different manner than filtering is done in Dudley's channel vocoder.

Linear predictive encoding of speech can be realized by a single complicated integrated circuit chip that performs 12 million multiply-adds a second. The 4.8-CELP (Code-Excited Linear Prediction) encoder used by the military encodes speech in a stream of 4800 binary digits a second, which can be transmitted over most dialed-up telephone circuits. The bit stream is encoded for secrecy. The quality of the reconstructed speech is less than telephone standard quality. The 4.8-CELP system will be put forward for commercial use in mobile telephone and for satellite-to-ground transmission of voice.

It is expected that an 8-CELP linear prediction system operating at 8000 bits per second will be standardized by 1992. This system is intended for use in cellular telephone (commercial mobile radio). It is hoped that it will increase the number of calls that can be transmitted in the assigned frequency band by a factor of seven.

In 1936 Homer Dudley showed that one need not transmit the speech waveform in order to transmit speech. Intelligible speech could be reconstructed by means of a talking machine, the vocoder, supplied with an apt description of the spoken sounds.

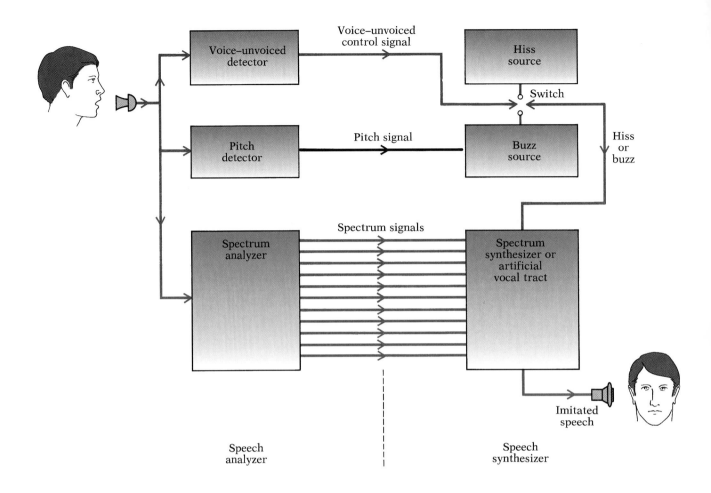

In human speech, a voiced sound, such as a vowel, causes the vocal cords to vibrate, producing a buzz. An unvoiced sound, such as s, sh, or th, does not and sounds more like a hiss. The vocoder sends a voiced-or-unvoiced signal, and if a sound is voiced, a pitch signal. The vocoder also determines how the shape of the vocal tract has emphasized various ranges of frequency, and produces a number of signals that describe the spectrum, or frequency distribution. At the receiving end, an unvoiced signal selects a source of hiss sound, and a voiced signal selects a source of buzz sound. A pitch signal adjusts the pitch of the buzz source. The hiss or buzz source goes to a spectrum synthesizer, or artificial vocal tract, which is controlled by the signals that describe the speech spectrum. The output of the spectrum synthesizer is an intelligible imitation of the original speech.

Simplicity behind Complexity

The linear predictive vocoder reduces the bit rate by a factor of around one-tenth. This rate is still greater than the rate needed to transmit speech motions. The idea of sending signals to drive a speaking machine is good, but in the encoding one doesn't really get all the way back from the rapidly changing waveform to the slower though complex motions of the tongue, jaws and velum, and to the muscular tenseness and the changes in air flow that produce vibrations of the vocal cords or the shushing of air past constrictions in fricative sounds.

There exists in fact a commercial device that does attain much greater reduction in bit rate over that necessary in encoding the sound directly into a bit stream. Both Boesendorf and Yamaha produce pianos that make a highly accurate digital recording of the motions of the keys and the pedals during a piano performance. In playback the piano keys and pedals are driven from the recorded digital signal, reproducing every nuance of the performing artist—but omitting audience noise.

A compact disc would record and replay the piano sound stereophonically at around 1,400,000 bits per second. This recording would reproduce the piano sound fairly well, but not as well as playing the piece on the original piano using the recording of the piano's motions.

In the Yamaha piano the record is made on a $3\frac{1}{2}$ inch floppy disk, which can store about 720,000 eight-bit bytes, or 5,760,000 bits. A fair number of these bytes store instructions for the computer that records from and plays the piano, so we will overestimate the bit rate used in playing the piano. The record on the disc can play the piano for about 90 minutes, so we compute the bit rate used in playing the piano to be about 1066 bits per second. That is less than a thousandth of the bit rate of the compact disc recording—an amazing saving.

What a pity that we can so rarely get behind the complexities of the sound to some simple description of how it can be produced—or reproduced. As an example of how we could get at the simplicity lying behind the seeming complexity of a picture rather than a sound, consider transmitting a marionette performance. We might use television. But how much less costly it would be to transmit the motions of the strings used by the puppeteer to control the marionette. For one marionette, I estimate the saving to be by a factor of about 50,000.

4
• • • • —

Modulation and Encoding

$\mathbb{A}$ll of us have encountered the word *modulation* in connection with FM (frequency modulation) and AM (amplitude modulation) radio. Modulation, and its inverse *demodulation*, make it possible to tune in on one radio or TV program when many are "on the air." *Encoding* is a familiar word that implies the representation of one thing by another. Modulation and encoding are essential elements of most communication. As we shall find in the next chapter, they make it possible to send many independent messages over one pair of wires or one optical fiber. Here in this chapter, we will discuss the basic principles that make such wonders possible.

Frequency Shifting

As we have seen, any signal can be represented as the sum of sine waves of different frequencies. The bandwidth of the signal is the range of frequencies of these sine waves. Bandwidth is a measure

Used for maritime communication, this earth station in Goonhilly, England, sends and receives signals from a British Telecom satellite.

The Frequency Range of Various Communication Services

ELECTRO-MAGNETIC SPECTRUM	FREQUENCY (CYCLES PER SECOND)	FREQUENCY RANGE	COMMUNICATION SERVICE
Radio	10		
	10^2	Extremely low frequency (ELF) Hertz	Telegraph, teletypewriter
	10^3	Voice frequency (VF) 300 Hz to 3 kHz	Telephone circuit
	10^4	Very low frequency (VLF) 3 to 30 kHz	High fidelity
	10^5	Low frequency (LF) 30 to 300 kHz	Fixed, maritime mobile, navigational radio broadcast
	10^6	Medium frequency (MF) 300 kHz to 3 MHz	Land and maritime mobile radio, radio broadcast
	10^7	High frequency (HF) 3 to 30 MHz	Fixed, mobile, maritime and aeronautical mobile, radio broadcast, amateur radio
	10^8	Very high frequency (VHF) 30 to 300 MHz	Fixed, mobile, maritime and aeronautical mobile, amateur radio, television broadcast, radio location and navigation, meteorological communication
	10^9	Ultrahigh frequency (UHF) 300 MHz to 3 GHz	Television, military, long-range radar
	10^{10}	Superhigh frequency (SHF) 3 to 30 GHz	Fixed, mobile, radio location and navigational, space and satellite communication, microwave systems
	10^{11}	Extremely high frequency (EHF) 30 to 300 GHz	Waveguides, radio astronomy, radar, radiometry
	10^{12}	Far infrared region 300 GHz to 3 THz	
Infrared	10^{13}	Mid-infrared region 3 to 30 THz	
	10^{14}	Near infrared region 30 to 400 THz	Optical fibers
Visible		Visible	Heliograph, signal flags

1 kilohertz = 1000 hertz; 1 megahertz = 1,000,000 (10^6) hertz; 1 gigahertz = 1,000,000,000 (10^9) hertz;
1 terahertz = 1,000,000,000,000 (10^{12}) hertz

of the frequency space. A signal can be shifted up or down in frequency by shifting the frequencies of its component sine waves.

Consider a signal with sinusoidal component frequencies extending from near zero to some upper limit, B. Such a signal is commonly called a baseband signal. It is called so partly, no doubt, because it is the basic signal that we start with, usually a sound wave but perhaps a video signal from a TV camera. Also, the baseband signal represents the signal in its lowest possible or most basic frequency range.

A signal can be shifted from its baseband range of frequencies to a higher frequency range without altering its amplitude and phase. A baseband signal with frequencies from near 0 to 4000 Hz, for example, might be shifted from 60,000 to 64,000 Hz (60 to 64 kHz). The bandwidths of the baseband signal and of the frequency-shifted signal are the same, namely, 4000 Hz (4 kHz). Shifting signals from their baseband range of frequencies to another, higher range of frequencies is accomplished through a process called modulation.

Modulation

Communication engineers do not ordinarily transmit signals in baseband form, except for local telephone service. There are several reasons.

Suppose we have a single pair of wires between two distant cities, and we wish to transmit more than one signal over the wire pair. If we transmitted one signal at baseband frequencies, we would use up the frequency range and could not transmit another baseband signal. In baseband transmission different signals must be sent over different pairs of wires, as occurs, for example, when a signal travels over the pair of telephone wires that goes to every office and to almost every home.

Consider radio. The wavelength of a radio signal is the velocity of light divided by frequency. Baseband signals have low frequencies and therefore long wavelengths. Antennas are most effective when they are a quarter of a wavelength long. At the highest telephone baseband frequency, 4 kHz, a wavelength would be 75,000 meters:

Velocity of light $= 3 \times 10^8$ meters per second

$$\text{Wavelength} = \frac{\text{velocity of light}}{\text{frequency}}$$

$$= \frac{3 \times 10^8 \text{ meters per second}}{4 \times 10^3 \text{ cycles per second}}$$

$$= 75{,}000 \text{ meters}$$

A quarter wavelength would be one-fourth of 75,000 meters, or 18,750 meters—a ridiculously long length for an antenna. Further, radio transmission at long wavelengths is plagued by noise during lightning and other disturbances. And just like transmission over wires, if we broadcast one signal at baseband frequencies, we would use up the frequency range and could not transmit another baseband signal.

Through modulation we can move baseband signals to different ranges of frequency. This allows us to send signals over wires or over the air in any of a large number of frequency ranges without overlapping.

Amplitude Modulation

Amplitude modulation (AM) is one way of transmitting a signal as a band of high-frequency components. This type of modulation is familiar to us in broadcast radio. An amplitude-modulated signal is a high-frequency wave whose peak amplitudes vary in accordance with an original, or baseband, signal. We shall see later how amplitude modulation is used to combine together a number of telephone signals.

In radio broadcast, the sound signal is first converted to a varying electric voltage by a microphone. The resulting baseband signal, whether for voice or music, varies from positive to nega-

An amplitude-modulated (AM) signal is a high-frequency wave whose peak amplitudes vary in accordance with an original, or baseband, signal. In radio broadcast, the voltage of the baseband voice or music signal varies from positive to negative (A). To convert the signal to AM, a constant voltage is added to the signal voltage, producing a voltage that is always positive (B). The sum of the constant voltage and the signal voltage is then multiplied by a high-frequency sine wave to produce the transmitted AM signal (C). The envelope, or outline of the peaks, of the modulated radio-frequency sine wave, called the carrier, reproduces the shape of the baseband audio wave, and so the amplitude-modulated carrier conveys the message of the original signal.

tive—that is, the amplitude of the wave sometimes rises above and sometimes dips below a voltage of zero. The first step in the conversion of the signal to AM is to add a constant voltage to the signal

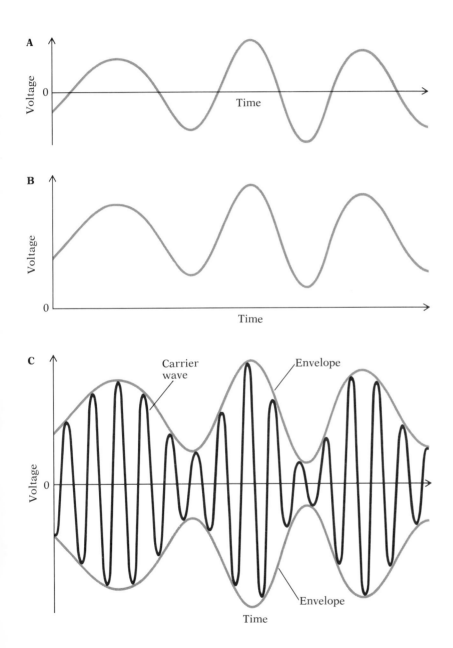

voltage to produce a voltage that is always positive. This allows the baseband signal to be recovered easily in the receiver.

Next, the sum of the constant voltage and the signal voltage is multiplied by a sine wave, called the *carrier*, lying in the radio-frequency range. That is, the values of the amplitudes of both waves at each and every corresponding instant in time are multiplied together. The device that produces the sine wave is an oscillator, and the device that multiplies the two signals is a modulator. We can view the new signal as an altered form of the carrier: its frequency is the same, but its maximum amplitudes, or peaks, now vary. After modulation, the outline of the peaks of the carrier, called the envelope of the wave, faithfully reproduces the baseband audio wave. The amplitude-modulated carrier conveys the message of the original voice or music signal.

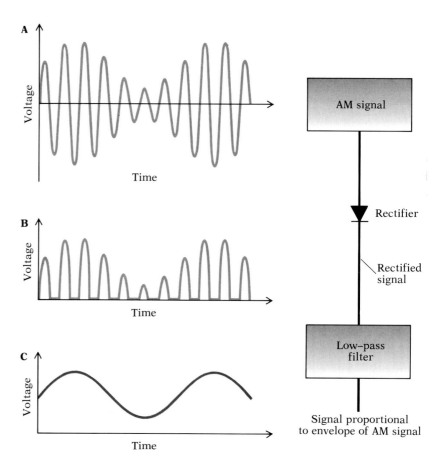

The original signal is extracted from the AM carrier wave in two stages. First the AM signal (A) enters a rectifier, or detector, which allows only the positive portions of the signal to pass (B). Next the rectified signal, consisting of high-frequency and low-frequency sinusoidal components, goes through a low-pass filter, which eliminates the high-frequency components. The remaining low-frequency components constitute a signal that is proportional in amplitude to the original signal (C).

CHAPTER 4

The original baseband signal can be recovered by a two-step process, called demodulation. The first step, called rectification, allows only positive portions of the amplitude-modulated carrier to pass. In the second step, low-pass filtering removes frequency components that lie above the original baseband. In effect, the demodulation process tracks the envelope of the amplitude-modulated carrier and thereby reproduces the baseband signal.

Sidebands

We now examine an amplitude-modulated wave from a new point of view. In the previous section, we thought of the signal as a single wave that changed with time. Now we consider the shape and width of the spectrum of the amplitude-modulated carrier. The signal itself and its spectrum are different representations of the same information.

As a result of the process of modulation, the bandwidth of the amplitude-modulated carrier, or received signal, turns out to be just twice the bandwidth of the original baseband signal. Two bands of frequencies are produced, one lying below and the other lying above the frequency of the carrier wave. These bands are called sidebands. Each sideband is equivalent in bandwidth and relative amplitude to the baseband signal; therefore, each sideband represents the entire information content of the baseband signal, shifted upward by the frequency of the carrier. In addition to the sidebands, there is a sinusoidal component equal in frequency to the frequency of the carrier wave; the carrier wave is like a catalyst and escapes unscathed in amplitude modulation. The upper sideband has the same identical shape as the frequency spectrum of the baseband signal, while the lower sideband has a mirrored shape.

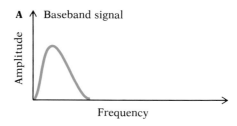

A. Baseband signal

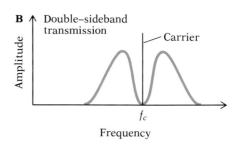

B. Double–sideband transmission — Carrier — f_c

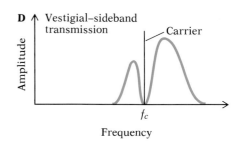

C. Single–sideband transmission

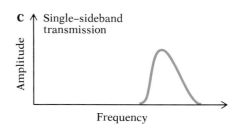

D. Vestigial–sideband transmission — Carrier — f_c

The frequency spectrum of an AM signal, which is nonperiodic, consists of a carrier-wave component and two sidebands. Each sideband is equivalent in bandwidth and relative amplitude to the baseband signal (A), shifted upward by an amount f_c. In double-sideband transmission (B), used in AM radio broadcast, the carrier and both sidebands are transmitted. Some forms of radio telephony make use of single-sideband transmission, in which only one sideband is sent and no carrier is transmitted (C). Television broadcast makes use of vestigial-sideband transmission, in which the upper sideband, the carrier, and part of the lower sideband are transmitted (D).

John R. Carson, a mathematician and pioneer in telephony at AT&T and later at Bell Laboratories, was the first to fully understand the process of modulation. Through his mathematical analysis in 1915, Carson showed that each sideband contains all the information in the original baseband signal. Thus the baseband signal can be recovered from either sideband. The carrier contains no information beyond the exact frequency and phase of the radio-frequency sine wave that the baseband signal modulated. The presence of the carrier, however, makes the recovery of the baseband signal simple.

In AM radio, the carrier and both sidebands are transmitted. This is called double-sideband transmission. The baseband signal is recovered simply by rectification and low-pass filtering. In radio telephony and in some other applications, only one sideband is transmitted (the other is filtered out), and no carrier is transmitted. Such transmission is called single-sideband (SSB) transmission. It conserves both transmitted power and bandwidth. A carrier of exactly the right frequency must be supplied at the receiver in order to recover the baseband signal exactly. In actual practice, it has been found that the carrier frequency used in demodulation can be in error by a few hertz without noticeably distorting voice signals.

In television transmission, the upper sideband, the carrier, and a part of the lower sideband are transmitted (a mode of transmission known as vestigial-sideband transmission). As a result, the radio-frequency bandwidth required to transmit the 4,200,000 Hz (4.2 MHz) baseband signal is reduced from 8.4 to 6 MHz. A slightly distorted but usable version of the baseband signal is recovered by rectification and low-pass filtering. Vestigial-sideband transmission uses somewhat more than half the bandwidth of regular AM. Single-sideband and vestigial transmission are advantageous because bandwidth and power are conserved, and transmission costs are reduced. Receiver complexity, however, is increased.

Frequency Modulation

In amplitude modulation, the variation in the maximum amplitude, or peaks, of the amplitude-modulated carrier can range from completely off to fully on, including all values in between. But suppose instead of changing the peak amplitude of the carrier, we change its frequency. We do so by making the frequency of the carrier signal, which has a fixed maximum peak and power, follow the changing voltage of the baseband signal. The frequency of the carrier increases as the amplitude of the baseband signal at each instant in time rises above a voltage of zero and decreases as the amplitude dips below a voltage of zero. This form of modulation is frequency modulation (FM), first demonstrated by Edwin H. Armstrong, a radio pioneer and an inventor. The diagram on this page illustrates one of several ways to accomplish frequency modulation.

When we use a baseband signal to modulate the frequency of a carrier wave, the amplitude of the baseband signal recovered at the receiver is proportional to the amount of frequency modulation, or the difference between the frequency of the unmodulated carrier and the frequency of the modulated carrier at any instant in time. This difference is called the frequency deviation. The maximum amount of frequency deviation occurs when the instantaneous amplitude of the baseband signal is at its maximum, and this amount of deviation is called the maximum frequency deviation. Increasing the maximum frequency deviation thus increases the maximum swings in frequency of the frequency-modulated carrier. The higher the maximum frequency deviation, the higher the signal-to-noise ratio in the received signal. The signal-to-noise ratio is improved by a factor of $(\Delta f_{max}/B)^2$, where Δf_{max} is the maximum frequency deviation and B is the bandwidth of the baseband signal. However, although increasing the maximum frequency deviation increases the signal-to-noise ratio in the recovered signal, it also increases the bandwidth.

A maximum frequency deviation many times greater than the bandwidth of the baseband signal increases the bandwidth necessary for transmission. This is called broadband FM. Unlike amplitude modulation, a frequency-modulated carrier can require a bandwidth many times the bandwidth of the baseband signal. In 1939, J. R. Carson proposed a rough rule of thumb: the bandwidth required to transmit an FM signal is twice the sum of the maxi-

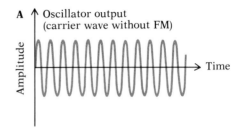

A Oscillator output (carrier wave without FM)

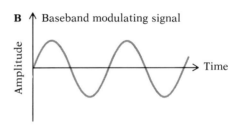

B Baseband modulating signal

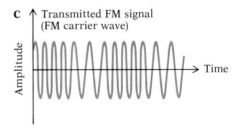

C Transmitted FM signal (FM carrier wave)

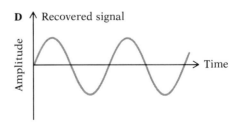

D Recovered signal

A frequency-modulated (FM) signal is a high-frequency sine wave, or carrier wave, whose instantaneous frequency varies in accordance with the instantaneous amplitude of a baseband signal. An oscillator produces a carrier wave that oscillates at a constant frequency (A). The arriving baseband signal (B) changes the frequency of the carrier wave by an amount proportional to the amplitude of the baseband signal (C). The baseband signal is recovered at the receiver (D).

mum frequency deviation and the bandwidth of the baseband signal, $2 \times (\Delta f_{max} + B)$. If the maximum frequency deviation is small compared to the maximum frequency component in the baseband signal, the bandwidth of the FM signal is simply twice the bandwidth of the baseband signal, just like AM. Such transmission, called narrowband FM, gives little or no improvement in signal-to-noise ratio compared to AM.

Frequency modulation is used for radio broadcast, for the sound channel of television, and for mobile cellular telephony. AM radio broadcast transmits a baseband signal with a bandwidth of 5 kHz over a radio-frequency bandwidth of 10 kHz. FM radio broadcast transmits a baseband signal with a bandwidth of 15 kHz over a radio-frequency bandwidth of 200 kHz. In (mono) FM radio broadcast, an improvement in signal-to-noise ratio of about 100 times, or 20 dB, is obtained—at the cost of a radio-frequency bandwidth about 20 times greater than that used in AM broadcast.

Increasing the bandwidth gives an improvement in signal-to-noise ratio that is consistent with Shannon's formula for channel capacity, which tells us that channel capacity should increase with bandwidth as well as with power. But Shannon's formula also tells us that increasing the bandwidth of transmission while holding the power constant can increase the signal-to-noise ratio only so much, as we now explain.

With a given transmitter power, we cannot obtain as large a signal-to-noise ratio as we wish simply by going to broader and broader band FM. We saw in the last chapter that noise power is proportional to bandwidth. As the bandwidth is increased, the bandwidth of the radio-frequency receiver must be likewise increased, and the noise mixed with the carrier sine wave of varying frequency will thus be increased.

As the distance between the transmitter and the receiver increases, the received signal becomes smaller. As the signal becomes smaller and smaller, the received signal-to-noise ratio does not degrade gradually but deteriorates suddenly. Deterioration is sudden because when the radio-frequency signal-to-noise ratio becomes too low, it is no longer possible to measure the frequency of the frequency-modulated radio wave. As a result, large bursts of noise called breaking occur relatively often in the baseband output. Breaking will be familiar to anyone who has listened to the bursts of noise in a faint, fading FM signal on an FM receiver in an automobile. Far from the transmitter, the received signal fades, the signal-to-noise ratio falls, and the radio signal becomes weaker than the noise.

Besides improving the signal-to-noise ratio of the recovered baseband signal, frequency modulation offers two more advantages. The radio-frequency signal has a constant peak amplitude, which means that we can operate the transmitter at full power and full efficiency all the time. In contrast, because power varies with maximum amplitude, an AM transmitter continually changes its power output. It is also easier to amplify an FM signal. An amplifier strengthens a signal by reproducing it at a higher voltage. If an amplifier is to reproduce accurately the continually changing amplitude of an AM or SSB signal, the radio-frequency output voltage of the amplifier must be exactly proportional to the radio-frequency input voltage. Such an amplifier is called a linear amplifier. A high degree of linearity is hard to attain, especially at radio frequencies, and linear radio-frequency amplifiers are more costly and less efficient than nonlinear amplifiers. But because FM signals have a constant peak amplitude, we do not need linearity for FM. In FM we need not preserve the amplitude of a signal but only the way its frequency varies with time.

A fourth advantage of FM is that when two interfering signals reach a receiver, the stronger captures the receiver so that the weaker is not heard. The capture phenomenon is important in mobile radio, for we want to hear the nearest and strongest transmitter without interference from more distant transmitters.

The microwave systems used to transmit telephone signals across the country use FM with only a narrow frequency deviation. Engineers have chosen to use frequency modulation rather than AM or SSB transmission at microwave radio frequencies merely to avoid the necessity for linear amplifiers. The frequency deviation is so small, however, that it protects no better against interfering transmitters than AM or SSB transmission. Indeed, as radio-frequency amplifiers have become more linear, single-sideband AM transmission has been used in place of narrow-deviation FM. By using SSB/AM transmission, a microwave system can transmit more than three times the number of voice circuits than it can using FM.

The TH microwave radio system, operating in the 6-gigahertz (6,000,000,000 Hz) radio band, carries telephone signals across the continent. In 1979, when the TH system used narrow-deviation FM, each radio channel carried 2400 voice circuits. Two years later, in 1981, that capacity was increased to 6000 voice circuits per radio channel through the use of single-sideband suppressed-carrier transmission. The single-sideband TH system, called AR6A, carries a total of 42,000 two-way voice telephone circuits, which

Microwave radio carries communication signals across continents from repeater station to repeater station. Many repeater stations are located in remote areas, such as this Canadian facility at the peak of Pyramid Mountain in the Alberta Rockies.

are transmitted over seven two-way radio channels in the 6-gigahertz band. Even so, the drive to pulse code modulation, or digital encoding, is so strong that even relatively new AR6A systems are being replaced by digital PCM microwave systems.

Some recent communication systems use SSB amplitude modulation to transmit telephone signals over communication satellites. SSB/AM conserves bandwidth compared to FM. Although highly linear amplifiers are needed at the earth stations, these are now available. Comstar IV initially used frequency modulation, which allowed 1800 voice signals per transponder. With the use of SSB/AM, the same Comstar IV carries 7800 voice signals per transponder, for a total capacity of 93,600 two-way voice circuits.

Pulse Code Modulation

In 1948, Shannon, Bernard M. Oliver (later a vice-president and board member of Hewlett Packard), and I, John Pierce, were so struck by the advantages of another form of modulation, called pulse code modulation (PCM), that we published a paper, "The Philosophy of PCM," in the *Proceedings of the Institute of Radio Engineers*. In PCM, the waveform of a signal is encoded as a stream of digits, which are transmitted as on-and-off pulses. Hence the term *digital* is also used to describe PCM. We expected PCM to sweep the field of communications, but its general acceptance took longer than we had anticipated. Only now with optical fiber and compact discs has our vision become reality.

The mathematical rule that is the basis of PCM is the sampling theorem, discussed in the previous chapter. According to the sampling theorem we can convey the entire message of an information signal by sending only the amplitudes of the signal at specific instants, called sampling times. If a baseband signal has frequency components lying totally within a bandwidth B, then it can be represented accurately by its amplitude at $2B$ points each second.

Consider, for instance, a telephone signal made up of sinusoidal components whose frequencies lie in the base bandwidth that extends from 0 to 4000 Hz. If we measure the voltage of this time-varying signal, or waveform, 8000 times per second, the measured amplitudes tell us all there is to know about the original waveform. Thus, we can reconstruct the original waveform from the succession of measured amplitudes.

To transmit a signal of bandwidth B using PCM, we sample its amplitude $2B$ times per second. We represent the amplitudes of the successive samples of the waveform approximately by binary numbers. The table in the margin shows the correspondence between binary numbers and decimal numbers. Decimal numbers are based on powers of 10, binary numbers on powers of 2. A 1 farthest to the right in a binary number represents one, a 1 second from the right two, a 1 third from the right four, a 1 fourth from the right eight, and so on, in powers of 2. Once in binary form, the numbers are transmitted as on-and-off pulses, as in teletypewriter and data transmission. Each sequence of pulses is a code for a sample amplitude, hence the name pulse code modulation.

In transmission, the binary numbers of a PCM signal can be represented in various ways—for example, by changes in the phase of a sine wave, by pulses with positive and negative amplitudes, or by the simple presence or absence of pulses. The representation is always equivalent to a succession of binary 0's and 1's. Even if some noise has been added to a binary PCM signal during transmission, it is possible to regenerate the PCM signal along the route before sending it farther. We regenerate the PCM signal by determining the original string of 0's and 1's and by then sending

Correspondence between Binary Numbers and Decimal Numbers

BINARY NUMBER	DECIMAL NUMBER
0000	0
0001	1
0010	2
0011	3
0100	4
0101	5
0110	6
0111	7
1000	8
1001	9
1010	10
1011	11
1100	12
1101	13
1110	14
1111	15

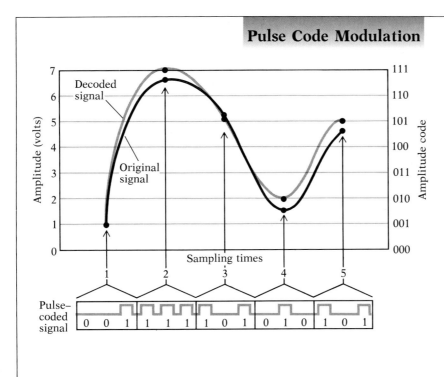

Pulse Code Modulation

The entire message of a signal waveform can be conveyed by transmitting only the amplitudes of the signal taken at specific sampling times. In pulse code modulation, sample amplitudes are represented as binary numbers, which are encoded and transmitted as sequences of off-and-on pulses. The original signal in the graph is part of a voice signal whose amplitude lies between 0 and 7 volts.

At appropriate sampling times 1, 2, 3, 4, and 5, the amplitudes of the signal are 1.0, 6.6, 5.2, 1.5, and 4.6 volts. Each sample amplitude is approximated to the nearest unit, and the approximate amplitude is represented by the corresponding number in a binary code, as shown in the following table:

out a nice, fresh, nonnoisy signal to the next amplifier or repeater. The determination of the original string is made by a simple threshold decision: if the received voltage is above some threshold, the binary digit is a 1; if it is below the threshold, the binary digit is a 0.

At the receiver we can reconstruct the original signal from sample amplitudes. We first produce a sequence of $2B$ short pulses per second whose amplitudes are proportional to the amplitudes of the original waveform (with maximum frequency B) specified by the received binary numbers. We then pass the sequence of pulses (called a pulse train) through a low-pass filter, which passes all the sinusoidal components of the pulse train whose frequencies lie below the frequency B and eliminates all the components of higher frequencies. The output of the filter is the original waveform.

AMPLITUDE	CODE
0	000
1	001
2	010
3	011
4	100
5	101
6	110
7	111

A three-digit binary number is transmitted as a sequence of three pulses (a 1 is represented by an on pulse and a 0 by an off pulse, or the absence of a pulse). The amplitudes of the five samples can be encoded as shown in the table below.

The pulse-coded signal is received with complete accuracy in spite of considerable noise. At the receiver, the signal represented by the code can be reconstructed perfectly, but the decoded signal is a little inaccurate. An error of as much as plus or minus half a unit of amplitude may occur in choosing the code that fits the amplitude as nearly as possible.

SAMPLE NUMBER	TRUE AMPLITUDE	APPROXIMATE AMPLITUDE	CODE
1	1.0	1	001
2	6.6	7	111
3	5.2	5	101
4	1.5	2	010
5	4.6	5	101

Transmission by PCM, like FM transmission, exhibits a breaking phenomenon. If the radio-frequency signal-to-noise ratio is inadequate, the receiver can make mistakes in interpreting the binary digits, and the recovered signal suffers huge errors.

Of course PCM cannot be better than is allowed by Shannon's formula for channel capacity. But how good is PCM? In PCM, the rate of rise of the signal-to-noise ratio is steeper than for FM, but for radio-frequency bandwidths typical of FM, the overall performance of FM and PCM are comparable. However, PCM has special advantages that make it pretty much the choice of the future.

In a PCM system, even if we receive without error all the binary digits that represent a sample, on the average the reconstructed signal will be in error because we must round off a sample amplitude to the closest binary number. To get a satisfactorily accurate result we must use a much more accurate representation of sample amplitudes than can be given by four-digit binary numbers. Usually 8 digits are used for telephone speech, and 16 for high-fidelity recording on compact discs. On compact discs, the binary digits are recorded as tiny pits of different reflectivity and are read by a beam of light from a semiconductor laser.

Some people talk loudly when they use the telephone and others talk softly. In using PCM for telephone speech signals, it is highly desirable to try to make the fractional error in representing samples by means of binary digits roughly the same for all sample amplitudes so that loud speech and soft speech will be reproduced with about the same error. Thus in actual PCM systems designed for telephony, the binary numbers that are transmitted do not give the amplitude of the voice signal directly. Rather, each binary number designates a particular voltage, and the voltages are closer together for low-level signals and farther apart for high-level signals. In other words, the voltages, or quantizing levels, are spaced in such a way that the distance between one voltage and the next is roughly proportional to the voltage. Therefore, the error in representing a voltage is roughly a constant fraction of the voltage. As a result, loud voices are represented with about the same fidelity as soft voices. In a compact disc, however, the voltages are all spaced equally apart with linear step sizes.

Like frequency modulation, PCM makes it possible to get a better signal-to-noise ratio by using more bandwidth. The more pulses (binary digits) we transmit per second, the more bandwidth we need to transmit them, but the more accurately we represent the signal. The bandwidth needed to transmit a PCM signal is roughly the number of binary digits used to represent each sample times the bandwidth of the baseband signal. Thus, if 8 bits are

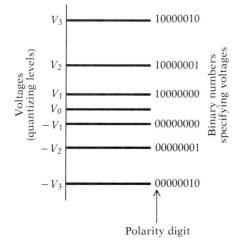

The binary numbers that are used in telephone transmission by pulse code modulation do not give the sample amplitude of the voice signal directly. Rather, each binary number specifies a particular voltage, or a quantizing level (here quantizing means rounding off). The voltage nearest that of the sample voltage to be transmitted is chosen, and the corresponding binary number is transmitted. Only the quantizing levels that are near zero voltage are shown in the illustration. One of the binary digits, called the polarity digit—the first in this instance—specifies whether the voltage is positive or negative (above or below zero voltage).

used for each sample of a telephone signal that has a bandwidth of 4 kHz, the bandwidth of the PCM signal is approximately 32 kHz.

Modulation is essential if we are to send a message by means of radio waves or light waves. But the frequency shifting characteristics of AM and FM and the digital transmission characteristics of PCM have another important use in communication. They are the basis of multiplexing, the technique that allows many signals to travel simultaneously over a single path or medium.

5
· · · · ·

Multiplexing: Many Messages on One Medium

When people first spoke between Boston and Chicago in 1893, a pair of heavy copper wires, each about a sixth of an inch in diameter, carried a single conversation. By 1915, telephony had spanned the continent, but, again, a single call monopolized a pair of heavy wires.

A pair of wires spanning the continent is an expensive investment. Long-distance telephony was bound to be costly as long as one pair of wires was needed for each call. When transcontinental service across the United States was inaugurated in 1915, the charge was $22.20 for a three-minute call, or about $260.00 in current dollars. Today, long-distance calls are not regarded as a luxury. Parents, children, other relatives, and friends expect long-distance calls—and receive and make them. Businesses count on reaching faraway contacts in an instant. Calls have become cheaper because today many people share the cost of a wire or other communication medium (such as optical fiber or radio) by

Cross-sections of the many different types of cables used to carry telephone circuits across town, across continents, and under oceans.

Telephone communication grew rapidly during its early years, resulting in the haphazard stringing of wires. During disasters, such as the blizzard of 1888 in New York City, many wires collapsed, disrupting service.

talking over it simultaneously. The sharing of a medium and its path is called multiplex transmission. The continual lowering of long-distance rates through advances in multiplex transmission has changed our national behavior and our sense of community as drastically as local telephony has changed our everyday lives.

Multiplex telephone transmission was foreshadowed by duplex and quadruplex telegraph transmission, ingenious and obsolete schemes quite different from the multiplex of today. It was also preceded by the use of phantom circuits in telephony, another obsolete art that gave three instead of two circuits on four wires, or seven circuits instead of four on eight wires.

Subsequently many conversations have been sent over a single communication path. Tens of voice signals can travel over a single pair of wires, and many thousands over one coaxial line. A coaxial line consists of a wire and insulating material in the center of a metal tube. The insulating material surrounds the wire, lying between the wire and the metal tube. In telephone jargon, a cable is many pairs of wires, or coaxials, bunched together, so telephone engineers call a single coaxial a pipe. In cable television, as many as 100 television signals are sent over one coaxial, except that CATV engineers call the coaxial a coaxial cable. It is no wonder that telephone people and TV people have trouble communicating with each other!

Some long-distance telephone signals are carried by microwaves (radio waves of short wavelength and very high frequency). Thousands of voice signals are sent over a single microwave path, whether the path is from hilltop to hilltop to hilltop or from earth to satellite to earth. In optical fiber, tens of thousands of voice signals are sent over a single strand of glass with a transmissive region that is about one-tenth the diameter of a human hair.

How on earth can different signals be kept from interfering with each other? Two types of multiplex transmission make it possible to send many conversations over one path. They are the older frequency-division multiplexing, which makes use of frequency

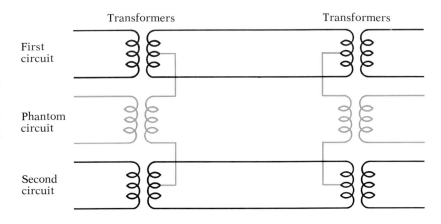

First circuit

Transformers

Phantom circuit

Second circuit

Transformers

The use of phantom circuits preceded successful multiplex telephony. Phantom circuits enabled two real circuits to do the work of three circuits, or four the work of seven. The diagram shows a phantom circuit (in color) derived from a pair of real circuits. A signal started off on the phantom circuit, then was shared between the two real circuits. A phantom circuit was connected to center taps on the transformers of the first and second circuits, isolating them from each other and preventing leakage, or crosstalk, between the real circuits and the phantom circuit.

shifting, and the newer time-division multiplexing, which makes use of digital transmission. This chapter explains the workings of these two types of multiplexing.

Filters

In frequency-division multiplexing, each separate signal in a group of signals is shifted to a different frequency range so that all the signals can share a common channel. The frequency shifting is accomplished by amplitude modulation, as described in Chapter 4. One of the steps in the process is to pass an amplitude-modulated wave through a filter to eliminate one of the sidebands. A filter responds to, or passes, some chosen range of frequencies and rejects all others. If we put a signal made up of sinusoidal components of many frequencies into the input of a filter, only the frequency components that the filter is designed to pass appear at the output.

Filters make use of the principle of resonance, which we observe, for example, in a child's swing and in a piano string. If you push a swing regularly at exactly the times that it swings fully back, it will swing even more strongly—that is, the motion will be reinforced. The swing is said to be resonant at this frequency. If you push the swing regularly at some other rate or frequency, it will not swing as strongly because the push will sometimes be against the direction in which the swing is moving. Or consider a piano string. If you sing a tone near the sounding board when the damper (loud) pedal is down, the tone will set a string of the same pitch into vibration.

In each case, the resonant system (swing or string) responds strongly to a frequency that is just right. There are many other sorts of resonators. Tuning forks, bells and gongs, organ pipes, the tubes under the wood bars of xylophones, and a small, closed room will all vibrate strongly in response to some frequencies and not to others. The old Tacoma Narrows bridge responded so strongly to winds at exactly the resonant frequency that the bridge swung violently enough to collapse. You can test resonance by singing at a variety of pitches in a small enclosure, such as a shower stall or empty closet, and listening for a resonant response. The shower stall will resonate when you hit a resonant pitch.

Swings and shower stalls are examples of mechanical resonators. When mechanical resonance happens, two factors are always

Thousands of telephone circuits are multiplexed together and carried over thin strands of optical fibers. The optical fibers are as thin as a human hair and fit through the eye of a needle.

CHAPTER 5

present: a restoring force, such as a spring or the pull of gravity, which tends to pull something back to center, and a mass or inertia, which tends to keep something that is moving from stopping. The stiffer a resonating spring, the higher the resonant frequency; the larger the mass, the lower the resonant frequency.

An electric filter is also a resonator, but it is the electric current that resonates. In an electric filter, the electric equivalent of a restoring force is a capacitor, which consists of two closely spaced electrodes, or sheets of metallic foil, and the electric equivalent of mass or inertia is the inductor, or coil of wire. Electric current flows through the inductor in one direction and then in the other, and so into and out of the capacitor. The box on page 48 explains in more detail how a combination of capacitors, inductors, and resistors act as a filter. By using a number of capacitors and inductors, it is possible to make low-pass filters that pass almost equally all frequencies below some chosen frequency and band-pass filters that pass almost equally all frequencies lying within some range, or bandwidth.

Low-pass filters are used in recovering the baseband signal from an AM or FM carrier wave, as was explained in the previous chapter. Band-pass filters are used in separating or selecting a single sideband of the two sidebands produced in the process of amplitude modulation. Band-pass filters in radio and television sets pass the signal from the desired radio or television station and reject signals from other stations.

The technology of communication advances when new physical phenomena are discovered and applied. Thus, the greatest advances in filters have come not from making better inductors and capacitors but from finding improved replacements for them. Piezoelectric materials, such as quartz crystal, provided one such replacement. These materials link mechanical motion with electric voltage. When they are stretched or bent, they produce a varying voltage. About 1917, Alexander McLean Nicolson, an engineer at Western Electric, realized that a quartz crystal with electrodes plated on it could act as a resonator in an electric oscillator and give a stable, accurate frequency. Crystal-controlled oscillators are used in communication systems, and also in electronic watches in place of balance wheels or tuning forks. Quartz crystals take the place of inductors and capacitors in filters.

For a time, the world was dependent on Brazil for high-quality quartz crystals. Since the 1950s, even better quartz crystals have been grown in the laboratory, and during the 1960s, synthetic quartz largely replaced natural quartz.

The resonant frequency of the thin, graceful Tacoma Narrows suspension bridge was easily excited by the winds that swept through the Narrows. This photograph captured the bridge swaying before it ultimately collapsed.

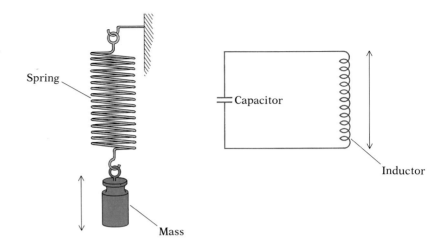

Resonance involves a restoring force, such as a spring, and a motion that tends to persist, such as that of a mass. A mechanical system (left) resonates at a frequency that depends on the stiffness of the spring and the size of the mass. In an electric circuit (right), the capacitor, which is made up of two closely spaced electrodes, is equivalent to a restoring force, and the inductor, or coil of wire, is equivalent to a mass.

Quartz crystals are a replacement for inductors and capacitors in filters. Another advance made it possible to replace only inductors, which are bulky, expensive, and less satisfactory than capacitors. Electric circuitry using transistors can simulate inductors, allowing filters (called active filters) made solely of transistors, capacitors, and resistors. Digital signals have spawned radically different filters. If a signal has been turned into digital form, as in PCM, a computerlike digital filter can accept the string of digits as an input signal and calculate an output string of digits that represents the original input signal passed through a filter. The function of the filter, be it conventional or digital, will always exist in communication systems. The way the function is performed changes and improves with time.

Frequency Shifting

The basis of frequency-division multiplexing is frequency shifting, a process that makes it possible to transmit different signals in different frequency bands. We saw in Chapter 4 that when a baseband signal is multiplied by a radio-frequency sine wave, or carrier, in the process of amplitude modulation, two sidebands are produced. Each reproduces the original baseband signal but is displaced upward in frequency by an amount equal to the carrier frequency. We can pass the amplitude-modulated carrier through a band-pass filter, eliminating one sideband and the carrier. The baseband signal is still shifted to a new frequency range, but the bandwidth is not changed. This kind of frequency-shifted signal is called a single-sideband (SSB) suppressed-carrier signal.

We can multiply the shifted signal by another sine wave and obtain two new sidebands. If we select one of these, the signal is shifted to yet another frequency range. The amplitudes and phases of all the frequency components are accurately preserved, and the original signal or the first frequency-shifted signal can be recovered by shifting all the frequencies down, using a sine wave of the appropriate frequency. The process of multiplying a signal by a sine wave and selecting one sideband is known as frequency shifting.

Frequency shifting has proved itself of service in many types of communication systems. In cable television systems, TV signals from a variety of sources such as over the air (VHF and UHF), tape machines, and local cameras are combined at the transmitting end of the cable system, called the head-end, and then sent over the cable. At the head-end, frequency shifting is used to move the TV signals to other frequency ranges (channel numbers) for convenient distribution via cable to the homes of the subscribers.

The microwave systems used in long-distance telephony, which send signals from hilltop to hilltop, must avoid interference

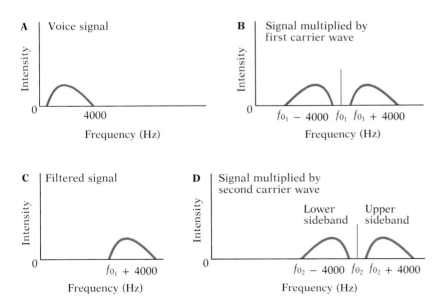

Frequency shifting moves signals to different frequency bands. When a voice signal with a bandwidth of 4000 Hz (A) is used to amplitude modulate a sine wave of frequency f_{01}, a lower sideband and an upper sideband are produced (B). Each reproduces the original voice-signal spectrum but is displaced upward in frequency. If a band-pass filter is used to select only the upper sideband of the signal (C), every sinusoidal component in the original voice signal is shifted upward in frequency by a frequency f_{01}. The shifted signal can be multiplied by another sine wave, and two new sidebands obtained. If one of these sidebands is selected, the signal is shifted by a frequency f_{02} to yet another frequency range (D).

between transmitter and receiver. To this end, a signal received in one frequency range is retransmitted at another by means of frequency shifting. To avoid interference, a communication satellite receives signals from earth in one frequency range, amplifies the signals, and retransmits them back to earth in another frequency range. Again frequency shifting makes retransmission possible.

In the superheterodyne radio receiver used in many home radio sets, the frequency of the received signal is shifted for convenience in amplifying it. A tunable radio-frequency oscillator produces a sine wave that is used to shift the desired radio signal to a fixed intermediate frequency range. In this way, each desired radio station is made to fall into the same frequency range, at which it is amplified by a high-gain amplifier of just the right bandwidth prior to detection.

Frequency-Division Multiplexing

By shifting the frequency of individual voice signals, frequency-division multiplexing produces a single signal of larger bandwidth that combines the individual signals. Each individual voice signal is carried over a single voice channel. Frequency-division multiplexing is one of the inventions that has made long-distance calls cheap enough for common use.

The process of frequency-division multiplexing begins when frequency shifting is used to shift 12 voice channels in a piece of apparatus called a channel bank. Each voice channel has a nominal bandwidth of 4 kHz, so that, after frequency shifting, the 12 channels lie one above another occupying a total bandwidth of 48 kHz in the frequency range between 60 and 108 kHz. Very good quartz filters select the sidebands produced in frequency shifting, so that each signal remains distinct from the others. But because even the best filters cannot be perfect, another measure serves to keep the voice signals separate. Although the nominal bandwidth allowed for one voice channel is 4000 Hz, only frequencies corresponding to the baseband audio range from 200 to 3400 Hz are actually present in the group of shifted frequencies. Therefore there are 800 Hz of unused frequency in each voice band. The unused frequencies between adjacent voice bands serve as a guard band to help prevent the signal from one band from crossing over into an adjacent band, a form of interference called crosstalk.

Each group of 12 channels can now be frequency shifted as though it were a single signal. Five groups are shifted into a 240-

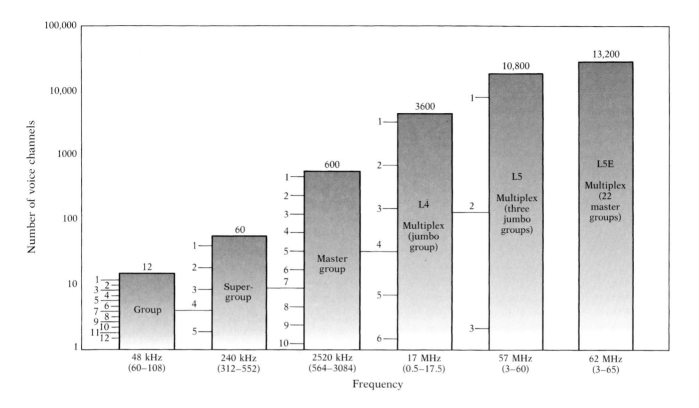

Frequency-division multiplexing produces composite signals of increasing bandwidth by shifting the frequency of individual voice channels. Initially, 12 voice channels (each with a nominal bandwidth of 4 kHz) are shifted in frequency so that they lie one above another in a "group" with a bandwidth of 48 kHz. Five groups are then assembled to form a "supergroup" of 60 voice channels, and 10 supergroups are assembled to form a "mastergroup" of 600 voice channels. In AT&T's L4 multiplex system, six mastergroups may be combined into one "jumbogroup" of 3600 voice channels, which are transmitted through a single coaxial-cable pipe. The L5 multiplex is able to transmit three jumbo groups in a single coaxial pipe. The L5 multiplex consists of coaxial pipes with a total system capacity of 108,000 two-way voice circuits. The L5E multiplex extended the capacity of the L5 system to a total capacity of 132,000 two-way voice circuits.

kHz-wide frequency band extending from 312 to 552 kHz; this forms what is called a supergroup. Frequency shifting can then be used to combine supergroups into signals of a greater frequency range, such as master groups and jumbo groups. In the end, we have a combined signal in which thousands of telephone channels lie cheek to jowl in ascending ranges of frequency, each telephone channel using a band of frequencies 4000 Hz wide.

Amplification and Negative Feedback

A frequency-division signal can travel thousands of miles across the country by coaxial cable or under the ocean by submarine cable. But it must be amplified every few miles or it will be attenuated to such a degree that it cannot be recovered. In the L5E coaxial-cable system, which can transmit 13,200 telephone channels through one coaxial-cable pipe, the signal is amplified every mile. In the TAT-7 submarine cable, which provides 4200 telephone channels under the Atlantic, the combined signal is amplified every 6 statute miles (about 10 kilometers).

An amplifier strengthens a signal by increasing its amplitude or voltage. It is essential that the amplifier for frequency-division multiplexed signals be linear—that is, the strength of the output signal must be strictly proportional to the strength of the input signal. If the output signal is amplified by a factor of two, it must be exactly twice as strong as the input signal at each and every instant in time. The output of a linear amplifier is an exact replica of the input, only larger. Yet amplifying elements (transistors or nearly obsolete vacuum tubes) are inherently nonlinear. Even slight nonlinearities in amplifiers can cause one telephone signal to interact with others, so that part of its energy is shifted to other frequencies. If that happens, the telephone signals cannot be separated cleanly at the other end. A signal in one channel will leak into other channels as unintelligible but noisy crosstalk.

It is also essential that the gain of the amplifiers be stable, maintaining as nearly as possible the same amplification at all times. Gain is the ratio of output power to input power and is measured in deciBels (abbreviated dB). In the L5 system, the combined frequency-division signal may be amplified several thousand times along the route, and the total amplification may be about a million deciBels (a power ratio of one to a one followed by 100,000 zeros). The slightest change in gain would make the received signal much too strong or much too weak. Frequency-division multiplex systems could not have developed as they have without the right kind of amplifier, one that is linear and that compensates for the problems of changes in gain. The amplifier that has been used successfully in frequency-division multiplexing is based on the principle of negative feedback. The negative-feedback amplifier is a highly ingenious and important device. Harold S. Black, a Bell Laboratories engineer, invented it at 8:15 A.M., August 2, 1927, while riding to work on the ferryboat from Hoboken, New Jersey, to Christopher Street in Manhattan. He was then 29 years old.

What is the secret of Black's negative-feedback amplifier? It is simply that Black used amplification, not to make the output as large as possible, but to make the output voltage as nearly as possible a specified multiple of the input voltage. In this way he was able to stabilize the gain of the amplifier. The multiple is determined by a part of the amplifier called the feedback path, which consists of resistors, inductors, and capacitors. The feedback path is linear and far more stable than the amplifying elements, whether they be vacuum tubes or transistors. And since the feedback path determines the multiple, the amplification is highly linear and stable.

If a fraction of the output of the amplifier is fed back along the feedback path and added to the incoming signal, positive feedback occurs. It increases the net input to the amplifier and thus increases the overall gain. But the amplifier can easily go out of control with the output seemingly increasing without bound until the actual amplifying elements finally overload. To overcome this problem, Black invented negative feedback.

If a fraction of the output of the amplifier is fed back along the feedback path to the input side and subtracted from the incoming signal, negative feedback occurs. It decreases the net input to the amplifier and thus decreases the overall gain and distortion. That seemingly simple fact is the basis of the far-reaching effects of the negative-feedback amplifier.

Although the fundamental idea is simple, it proved hard to build an amplifier with a large amount of feedback that did not go out of control and break into oscillation. The solution came from two Bell Laboratories mathematicians: Harry Nyquist, who developed a criterion for the stability of a feedback amplifier, and Hendrik W. Bode, who discovered a relation between phase and gain and showed how Nyquist's criterion could best be met.

Frequency shifting can be used not only to move signals up in frequency but also down in frequency. Thus, once a frequency-division multiplexed signal reaches its destination, frequency shifting can be used to extract supergroups, to extract groups from each supergroup, and finally to extract individual channels from groups. In this way, each telephone signal is returned to its origi-

The Negative-Feedback Amplifier

Transmission by means of frequency-division multiplexing requires an amplifier that is linear and stable. The negative-feedback amplifier is both: it produces an output signal that is strictly proportional to the input signal, and it maintains a constant amplification.

In an amplifier without feedback, changes in output are proportional to changes in voltage gain, which is the ratio of output voltage to input voltage. In an amplifier with feedback, a small amount of the output signal is fed back and mixed with the input signal. The advantage of this arrangement becomes apparent as the inherent voltage gain of the system is increased.

Where μ represents the inherent gain (the gain without feedback) and β the feedback fraction of the output,

Output voltage =
$$\mu\,[(\text{input voltage}) - \beta\,(\text{output voltage})]$$

$$\frac{\text{Output voltage}}{\text{Input voltage}} = \frac{\mu}{1 + \mu\beta}$$
$$= \frac{1}{\beta + 1/\mu}$$

Hence, if the gain μ is very large, the output voltage of the amplifier becomes almost entirely dependent on the fraction of the output that is fed back. If, for example, the inherent gain of an amplifier is 1000 times and the fraction fed back is just under one-tenth of the output, the actual net gain will be 10. The inherent gain in this system may drop to as low as 500, however, without materially changing the net gain. Thus, the actual net gain is set not by the amplifier's vacuum tubes and transistors but by its linear and stable components—the resistors, inductors, and capacitors that make up the feedback loop.

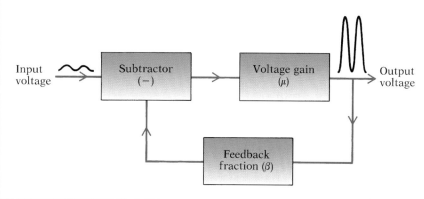

nal baseband frequency range on a separate pair of wires. If the amplification along the way is nearly linear, as it can be using Black's negative-feedback amplifier, the crosstalk between channels will be almost negligible.

Time-Division Multiplexing

Frequency-division multiplexing has served telephony well. It has made long-distance calls far cheaper, and it will continue to play a part in all electric communication. But another kind of multi-plex transmission, time-division, is both more economical and more satisfactory for short-distance transmission over wires and for all transmission over optical fiber, and it is also being used for long-distance communication over all transmission media. Time-division multiplexing works in an entirely different way from frequency-division multiplexing. In time-division multiplexing, several different signals are transmitted over a single channel, each signal going out at a different time.

We saw in Chapter 4 that a baseband signal with a bandwidth B can be represented by $2B$ successive sample amplitudes per second. These sample amplitudes can then be encoded as binary digits. The original baseband signal thereby becomes a PCM (pulse code modulation) signal.

Almost all the signals transmitted by time-division multiplexing are PCM signals. Suppose, for instance, that 8 on-or-off pulses represent a single sample of the amplitude of one voice signal at an instant in time. If the (nominal) bandwidth of the voice signal is 4000 Hz, then 8000 groupings of 8 pulses must be transmitted each second. That is, 64,000 on-or-off pulses per second must be transmitted in sending each voice signal. If we want to transmit 24 telephone conversations over one communication circuit, we must send at least $24 \times 64,000$, or $1,536,000$, pulses per second.

The diagram illustrates how the signals from many conversations share the same communication circuit. In the 24-channel time-division multiplexing system, 24 telephone lines lead to an electronic switch (switch 1 at left). Switch 1 samples the voltage in one line and relays the information to an encoder. The encoder translates the voltage value into a sequence of binary digits, which are transmitted as eight on-or-off pulses. The switch then "steps"

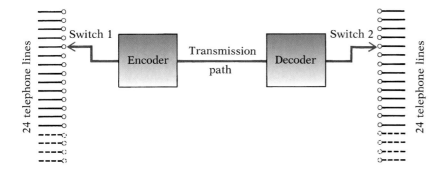

In a 24-channel time-division multiplexing system, the voltage of each line is sampled, encoded as a binary number, transmitted as eight on-or-off pulses, then decoded at the receiving end and connected to the correct telephone line. When all 24 lines have been sampled, the process begins all over again with the first line.

onto the next line, and the entire process is repeated. When all 24 telephone lines have been sampled, the switch returns to the first line and starts again; each of the 24 lines is sampled 8000 times per second. At the receiving end, a decoder reproduces the sample voltage represented by the eight on-or-off pulses. The decoded voltage is finally connected to the correct telephone line by a second electronic switch (switch 2), which is synchronized with switch 1.

Why has time-division multiplexing proved to be more effective than frequency-division multiplexing in reducing costs of transmission? In part because the cost of the equipment that does the multiplexing and demultiplexing is low. A frequency-division transmitting or receiving terminal is costly. Each channel requires a high-quality quartz band-pass filter, a constant frequency source, and a modulator. By contrast, most of the components in a time-division multiplex terminal serve all channels. At the transmitter, the same encoder can be shared to encode sample voltages from all channels as groupings of eight on-or-off pulses, and at the receiving end a common decoder can decode groupings of eight binary pulses to the corresponding sample amplitude. However, each channel does require a separate but simple low-pass filter to ensure that no frequency components higher than 4000 Hz are present before beginning the digital to analog conversion process. An electronic switch for sampling the waveform 8000 times per second is also needed.

Further, many of the functions in time-division multiplexing are binary, or on-off, in nature and can be performed by digital circuits. Thus, the development of complicated and inexpensive integrated circuits—or chips, as they are frequently called—has favored PCM and time-division multiplexing. Integrated circuits are not as well suited to the modulating and filtering functions needed in frequency-division multiplexing.

Pulse code modulation and time-division multiplexing prevent the accumulation of distortion. Devices called regenerative repeaters recreate clean PCM signals from stage to stage along the route. All they need do is determine whether an "off" or an "on" has been transmitted by simply examining whether each pulse is above or below a certain threshold. Then a clean, new PCM signal can be regenerated and sent to the next stage along the route. In contrast, the amplitude of an AM or SSB signal must be amplified exactly along a route, and any distortions that occur along the way will accumulate.

In a very long multiplex system, the cost of the circuit used for transmission is high, and the cost of the multiplexing terminal equipment is a small fraction of the total cost. In a very short system, the reverse is true. The cost of transmission is low, and terminal cost dominates. The first commercial PCM time-division multiplex system to succeed was the T1 system, which was put into commercial operation in 1962. Designed for distances from 5 to 100 miles, the system is used heavily on trunks between central offices. It multiplexes together 24 voice circuits, which are transmitted at an overall rate of 1.544 megabits per second. The system proves less costly than competing frequency-division systems because the terminal cost is lower. Further, transmission is stabler and its quality better than that provided by cheap frequency-division systems, in which corners must be cut to save cost.

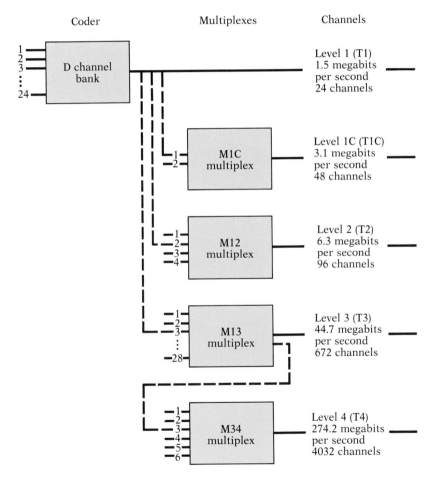

Coder   Multiplexes   Channels

D channel bank
1
2
3
⋮
24

Level 1 (T1)
1.5 megabits per second
24 channels

M1C multiplex
1
2

Level 1C (T1C)
3.1 megabits per second
48 channels

M12 multiplex
1
2
3
4

Level 2 (T2)
6.3 megabits per second
96 channels

M13 multiplex
1
2
3
⋮
28

Level 3 (T3)
44.7 megabits per second
672 channels

M34 multiplex
1
2
3
4
5
6

Level 4 (T4)
274.2 megabits per second
4032 channels

Time-division multiplexing interleaves the binary digits from many conversations. A D-type channel bank encodes 24 voice channels into a stream of 1.5 megabits (actually 1.544 million bits) per second, called a DS1 signal, which is carried over a T1 line. These megabit signals can take various paths through the multiplexing hierarchy. For example, two 1.5-megabit DS1 signals can be fed into an M1C multiplex and interleaved to produce a signal of 3.152 megabits per second. (Output signals exceed simple multiples of input signals because the multiplex adds several "housekeeping" bits.) Other multiplexes produce signals of even higher capacity, culminating in the M34 multiplex, the highest-capacity multiplex currently in use.

MULTIPLEXING

Time-division multiplexing is also being used increasingly for the transmission of voice signals over the local loop. The telephone company sometimes runs out of cable pairs in an area of large growth. By enlarging the capacity of existing cable pairs in the subscriber loop plant, multiplexing makes the installation of costly new cable unnecessary. Multiplexing in the local loop is called subscriber loop carrier. AT&T's SLC^R 96 system combines 96 voice circuits by time-division multiplexing and then sends the multiplexed signal over eight twisted pairs of copper wire or over optical fiber.

Does the advent of time-division multiplexing mean the end of analog modulation? Not at all. The signals multiplexed together create a new signal: the multiplex signal. The multiplex signal is transmitted over some communication medium, such as twisted pairs of copper wire, coaxial cable, radio airwaves, or optical fiber. This transmission usually requires an additional modulation. To create the transmitted signal, the multiplexed signal modulates a carrier wave, usually a sine wave.

In some cases, a time-division multiplex (TDM) signal is sent directly over a transmission medium, but in most cases a time-division multiplex signal also must modulate a carrier, usually a sine wave. A TDM signal modulates the peak amplitudes of a sine-wave carrier by simply turning the carrier on or off or by changing the peak amplitude from one value to another. A TDM signal can also modulate the sine-wave carrier by changing its frequency from one value to another. Finally, a TDM signal can change the phase of the sine-wave carrier. These modulation techniques can be used in combination: information travels more efficiently through simultaneous amplitude and phase modulation of a sine-wave carrier.

Thus, even though telephone communication is increasingly becoming digital through the use of PCM and time-division multiplexing, the ultimate transmission of the digital signal still involves analog modulation of a carrier wave. For example, the time-division multiplexed signal carried digitally over optical fiber modulates the amplitude of the light beam by turning it on and off. More than one channel can be carried over the same fiber by modulating different light beams at different wavelengths, a technique called wavelength-division multiplexing.

Time-Division Multiplexing and Satellites

The success of the *Echo* communication satellite in 1960, of *Telstar* in 1962, of *Syncom*, the first geosynchronous communication satel-

Communication satellites are frequently the only means of reaching remote regions. This earth station in Conakry, Guinea, is set up for satellite communication.

lite, in 1963, and the inauguration of commercial communication satellite service in 1965 demonstrate that space is really good for something—telecommunication. Although most long-distance calls travel through optical fiber, satellite links still connect remote areas, such as Canada's Northwest Territories, and developing nations scattered over the globe to the telephone networks of industrialized nations. Satellites also supply television to isolated homes in remote areas. They provide mobile service to ships and even trucks, and in the future, we hope, transoceanic telephone service to airplanes.

Most present communication satellites transmit broadband FM microwave signals. Each such FM radio channel transmits a signal made up of thousands of voice signals combined by frequency-division multiplexing. This is fine as long as the satellite must amplify and retransmit just one FM signal from one earth station. Suppose, however, that we want to use one satellite amplifier to amplify signals from several earth stations. One FM signal has a constant maximum amplitude, or envelope, and is not adversely affected by nonlinearities in an amplifier. The sum of two or more FM signals of different frequency ranges is not constant in maximum amplitude, however. Thus, if several FM signals are combined and passed through a nonlinear amplifier, they interfere with one another, and the recovered voice signals sound noisy. Hence, FM is not well suited for use in a system in which a satellite amplifier must amplify signals received simultaneously from many earth stations.

Time-division multiplexing is well suited for use in such a multiple-access system. Because each earth station uses the amplifier at a different time, the signals from the different earth stations cannot interfere with one another—unless they are sent at the wrong times and overlap when they reach the satellite. Such use of a satellite communication system is called time-division multiple access, or TDMA. The earth stations must synchronize their transmissions to avoid overlapping each other, or else they must simply take a chance in transmitting and then back off if interference occurs.

A satellite orbits 22,300 miles, or about 36,000 kilometers, above the nearest part of the earth's surface, and it is more than 26,000 miles from the remotest parts of the hemisphere from which it is visible. Thus, radio waves, traveling at the speed of

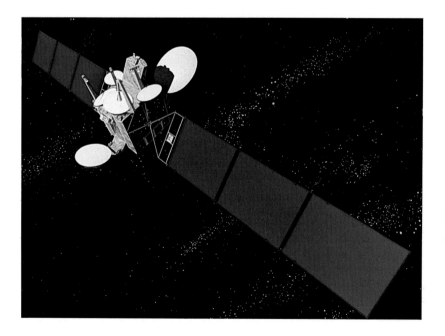

Completing one orbit in exactly 24 hours, communication satellites appear stationary relative to the earth turning below. A communication satellite, like the Intelsat VII, acts as a microwave repeater when transmitting signals.

CHAPTER 5

light, take one-eighth of a second to reach the satellite from a transmitting station directly under it and one-fiftieth of a second longer to reach the satellite from more distant parts of the earth's surface. Hence, it is necessary to ensure that signals transmitted from different earth stations do not wander from the time intervals assigned to them. But this problem of synchronization can be solved. It seems clear that in the future, communication-satellite systems will use PCM and time-division multiplexing rather than FM and frequency-division multiplexing.

The Future of Multiplexing Systems

Pulse code modulation and time-division multiplexing are well suited to terrestrial microwave systems when many transmitters and receivers must operate in the same area. In a PCM system, perfect operation means that one binary pulse can be distinguished from another. That is, the signal need only be good enough for the receiver to tell "off" from "on." If an AM or SSB system is used, the received signal must be good enough to permit recovery of the exact amplitude of the transmitted signal. An interfering transmitter strong enough to render an AM or SSB or narrow-deviation FM signal useless may not interfere at all with a PCM system. Because PCM is not sensitive to interference, digital transmission allows the use of more radio channels in a congested area, even though PCM requires more radio-frequency bandwidth per telephone circuit.

Because SSB/AM and narrow-deviation FM transmission tend to interfere with other systems operating on the same frequencies, long-haul microwave routes have been spaced far enough apart so that the signals from the transmitters of one route do not reach the receivers of another route. Once SSB/AM or narrow-deviation FM transmission is in use along a route, we cannot put more systems in the area by using PCM. A new PCM system would interfere with a nearby existing system. But digital transmission can operate near older systems if it exploits a new range of frequencies.

All common-carrier communication organizations are engaged in a transition from narrow-band frequency-division multiplexing to broadband time-division multiplexing of voice and data circuits. In part, this is a transition from older media—wires, coaxial cables, and terrestrial and satellite microwave systems—to the dominating medium of the future, optical fiber. In part, time-division multiplexing forms a natural marriage with the com-

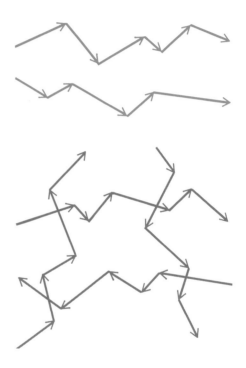

Microwave radio systems using single-sideband or narrow-band FM transmission are very sensitive to interference, and so identical frequencies can only be used over widely separated routes (top). In contrast, microwave systems using pulse code modulation can operate on the same frequencies along intersecting routes (bottom).

puter-controlled digital switching systems that are coming to dominate local switching and already dominate long-distance switching. The transition to broadband time-division multiplexing is inevitable, but it will be in some degree painful and costly because so much good frequency-division equipment is in service. The problem today is not deciding the direction to take but rather determining how to get to the destination most economically.

In 1988, AT&T wrote off all its analog frequency-division multiplex equipment in preparation for a faster transition to an all digital PCM network. In fact, AT&T's network is virtually all digital today with the exception of low-density routes and the maintenance of some alternate route capacity. Analog, if available, is used only as a last resort when all the digital circuits are occupied.

The transition of terrestrial microwave systems to time-division multiplexing is underway. The AT&T DR6-30-135 digital microwave system uses time-division multiplexing to transmit 2016 digital voice circuits per radio channel in the 6-gigahertz radio band. With seven active radio channels in that band, the total capacity of the system is 14,112 two-way voice circuits. This is far short of the 42,000 two-way voice circuits that are possible with the single-sideband AR6A system using frequency-division multiplexing. Indeed, time-division multiplexing is not as bandwidth efficient as frequency-division multiplexing.

There are sure to be more and more time-division multiplexing systems, but frequency-division multiplexing will not disappear entirely. Different radio signals will continue to be transmitted in different frequency bands, as in over-the-air broadcast radio and television and in cable television. However, the future of telephone communication lies with optical fibers, and PCM is ideally suited for optical fibers. Telephone signals are combined together by PCM and time-division multiplexing, but the actual transmission of the multiplexed signal by radio and some other means still involves the modulation of a carrier wave by AM, SSB, or FM techniques.

Converting a Signal from Two Wires to Four Wires

The telephone instrument in your home is connected to a single pair of wires, called the subscriber loop or local loop, which carries both the outgoing voice signal and the incoming one. The pair of wires creates an electric circuit for each of the two signals. A device in your telephone instrument called a hybrid, or hybrid coil, keeps the two signals separate, more or less, so that what you say into your telephone transmitter doesn't blast into your ear from the receiver.

In contrast, all multiplex systems provide separate talking paths in two directions. Separate paths are necessary because the amplifiers placed along the lines between terminals amplify sig-

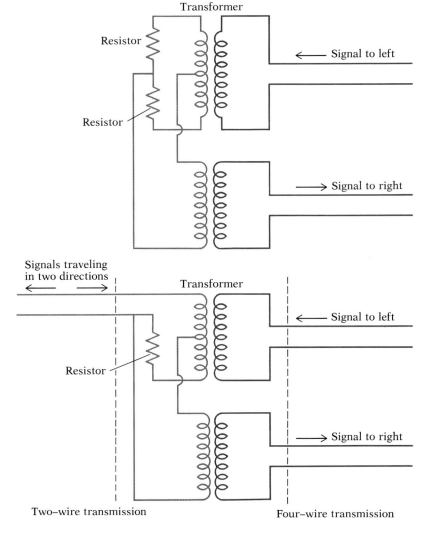

Top: In two-wire transmission, hybrids keep incoming and outgoing signals separate. When a signal comes in from the left, the voltage at the center tap of the transformer will be equal to that at the midpoint of the two resistors, preventing a signal from going out to the right. Bottom: Hybrids are used for conversion from two-wire transmission to four-wire transmission. If the two-wire line and the resistor are of equal resistance, a signal coming into the hybrid via four-wire transmission will leave via two-wire transmission.

All long-distance calls require the use of a hybrid to make the two-wire circuits that are connected to home telephones compatible with the four-wire connections of multiplex systems. The multiplex terminal interleaves signals from many four-wire voice circuits into one four-wire transmission path.

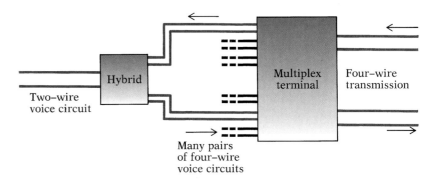

Two–wire voice circuit

Hybrid

Many pairs of four–wire voice circuits

Multiplex terminal

Four–wire transmission

nals traveling in one direction only. When two people talk between New York and San Francisco, or between any two distant cities, the call goes from one telephone through a local two-wire voice circuit to a multiplex terminal. There the call is transferred to a four-wire long-distance circuit that consists of two separate one-way voice circuits. At each end of the system, a hybrid reconverts each four-wire voice circuit into a two-wire circuit.

Four-wire transmission makes it possible to amplify a signal stably over a long distance. Two-wire transmission dates from the early days of telephony, before amplification. Because only one pair of wires is needed per conversation, it is inexpensive. And two-wire transmission is universal in local telephone transmission. Alas, two-wire transmission leads to the serious problem of echoes in long-distance calls. At the far end, where the four-wire long-distance transmission system is connected to the two-wire circuit, part of your voice signal is reflected back because the hybrid cannot be perfect and because electrical discontinuities exist along the local loop. To get rid of echoes, echo suppressors are used. They interrupt an incoming signal when someone is speaking and thus prevent the echo from returning back down the circuit. Unfortunately, they also prevent the other party's speech from returning, leading to choppy speech if both persons try to speak at the same time. An improved solution is the echo canceler. This device subtracts the echo from the return path through the use of special filters that can adapt themselves to the changing electrical characteristics of different two-wire circuits.

All long-distance transmission systems solve the problem of echo, whether through the use of echo suppressors or echo cancelers. Communication-satellite transmission has the additional problem of delay, because the time needed for the signal to travel from the earth to the satellite and back is so long. When conversing via satellite, you sometimes start to speak after waiting half a second for a reply and interrupt the very answer you sought.

Echo suppressors are satisfactory when the transmission time is moderate. They do not work nearly as well in communication-satellite circuits because the echo of your voice comes back delayed more than half a second after you speak. Then echo cancelers are the solution, but even using echo cancelers, satellite circuits to remote places like Sri Lanka can have unpleasant echoes.

I look to a day when local two-wire transmission will be replaced by four-wire PCM transmission by means of light waves over an optical fiber. But the replacement and modification of all the local telephone equipment will take a lot of money and a long time. Two-wire transmission and frequency-division multiplexing will be with us for years. In some forms, such as radio transmission, frequency-division multiplexing will be with us always. Time-division multiplexing, however, is the wave of the future, for satellite circuits as well as for terrestrial circuits.

6
— • • • •

Electricity and Light as Waves: Signals as Physical Phenomena

Something that is still missing from a complete picture of signals is an explanation of their physical nature, and of how they actually travel from one place to another. We have characterized signals in terms of frequency and bandwidth. We have seen how signals can be encoded, how many signals can be combined for common transmission by wires, by optical fibers, by radio waves. As for the signals themselves, we have thought of them as "electric." We have used the terms *current* and *voltage* and have spoken of current and voltage as changing with time.

How do electric signals get from one place to another? It is true, but not enough, to say that we now know that such signals can travel along wires, through air and space, through optical fibers. Are there different sorts of electric signals, or are all such signals the same in nature? With what speed do they travel? Do they become changed or distorted in transit? Our view of commu-

A wave is a disturbance that travels away from its source. Here the wave is in the form of ripples on the surface of a pond.

nication would be incomplete if we did not address such questions concerning the physical nature and behavior of communication signals.

The Signal as an Electric Current

Morse and Bell knew that if a battery is connected to a closed electric circuit, an electric current will flow in the circuit. In Morse's case the circuit consisted of a long wire, the electromagnet of a telegraph sounder, and a return path through the ground, which acted as a conductor. Thus, the current flowed from one terminal of the battery through the wire, through the electromagnet, to a metal stake driven into the ground (a "ground" electrode), back through the earth to another stake, and thence to the other terminal of the battery. In later communication, the ground return path was replaced by a second wire.

The picture of a signal as an electric current that flows all the way to the end of a closed circuit and back is appealingly simple, and it is correct if the current is never turned on or off. The early telegraphers ran into problems when they attempted to turn the current on and off rapidly. In some cases the current appeared not to get to the end of the circuit. In others, the current became distorted along the way. The early understanding of a signal as an electric current that was the same all along a circuit simply did not go far enough. In order to transmit electric signals over great distances, people had to have a fuller understanding of the nature of electric signals—of their velocity through air and other solids, and of the relation between the wavelength of a signal and the length of a transmission line.

Morse became aware of such problems of transmission when he started to put his first long telegraph line (from Washington to Boston) underground. The farther he went, the poorer the line performed. Morse simply abandoned the underground line, choosing instead to string his wires between insulators on poles.

The problem that Morse had encountered came up again when people first tried to telegraph across the ocean by means of an insulated wire placed on the ocean bottom. When the first transatlantic cable began carrying signals in 1858, the problem of delayed, distorted signals became glaringly apparent. Signaling was excessively slow. When the telegraph key was closed, the current at the far end of the circuit did not rise immediately; it crept up slowly. When the sending key was opened, the current at the far

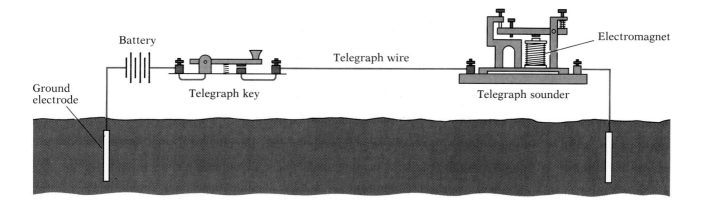

Battery

Ground
electrode

Telegraph key

Telegraph wire

Telegraph sounder

Electromagnet

end did not cease immediately; it fell gradually. The first cable failed in four weeks. But the slowness of signaling persisted even in the first successful cable, which was laid in 1866.

The English physicist William Thomson, later Lord Kelvin, analyzed the problem and explained the difficulty. What happens in an electric line is almost exactly analogous to what happens when a flame is applied to one end of a long metal rod; the far end does not get hot immediately. Moreover, if we take the flame away after the other end has become hot, that end does not get cold immediately. Just as heat can be stored in a metal rod, electric charge can be stored in the capacitance (charge-storing capacity) formed by the insulated wire of the cable and the surrounding water. The charge comes from the current that flows into the cable. No appreciable current will reach the far end until electric charge has accumulated all along the cable.

Just as heat flows slowly through a metal rod, so the electric resistance of a cable wire inhibits the electric current, the flow of electric charge through the wire. If the cable wire had negligible electric resistance, the signal that first arrived at the receiving end would be nearly identical in voltage to the signal transmitted. In practice, however, when a voltage is applied to one end of the cable wire, current flows into the wire, charge accumulates on the wire, and the voltage rises along the wire only when sufficient charge has accumulated. The higher the resistance per unit length, the more slowly charge accumulates and the more slowly the voltage rises.

The speed at which the voltage rises depends also on the capacitance of the cable, its charge-storing capacity. The larger the

The early view of the transmission of telegraph signals was simple. When the telegraph key was pressed down, the circuit connected to the battery was closed, and an electric current flowed through the circuit. The circuit consisted of the telegraph key, the wire coil of the electromagnet in the telegraph sounder, and the return path through the earth between the two ground electrodes.

The first successful transatlantic cable was laid by the Great Eastern, *the largest ship then in existence, in 1865. Today, the laying of undersea cables could be well described as routine—routine with infinite care.*

amount of charge that can be stored, the more slowly the voltage rises. The wire and the ocean (or ground) are two conducting surfaces separated by an insulator; hence, as we learned in Chapter 2, they form a capacitor. The capacitance of the cable can be increased by increasing the surface area of the wire or by decreasing the separation between the wire and water.

Kelvin's analysis of the flow of current through an electric cable was a marvelous advance. It was accurate enough for all practical purposes. Similarly, the idea that a current flows instantly all along the circuit had been accurate enough for a short overhead telegraph circuit. But Kelvin's analysis was not complete because it did not explain the behavior of all electric signals. In

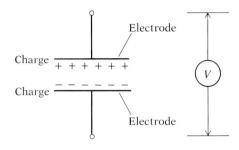

A capacitor is an arrangement of a pair of conductors (electrodes) that are separated by an insulator. When a voltage is applied to a capacitor, charge accumulates on the electrodes, and an electric field is produced. The symbol for a capacitor is two parallel lines, which represent two electrodes spaced a small distance apart.

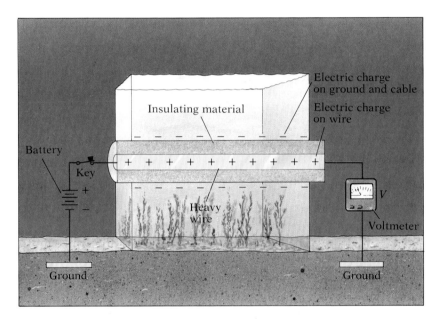

When an insulated wire is buried for use as an underground or underseas cable and a voltage is applied to one end, a voltage does not appear immediately at the other end. The delay in voltage is caused by the electric resistance and the capacitance of the cable. Current flowing through the cable first accumulates as electric charge between the cable and the surrounding ground or seabed, and a voltage arises at the far end only when enough electric charge has flowed into the cable to raise the voltage along its entire length.

1873 the Scottish physicist James Clerk Maxwell published a book that did, his *Treatise on Electricity and Magnetism*. Maxwell's book, and particularly his equations describing electric and magnetic phenomena, are a marvel of nineteenth-century science, and of all science, for that matter. I, John Pierce, remember seeing a play called *Wings Over Europe* more than 50 years ago, when I was a student at Caltech. The scientist hero died murmuring Maxwell's equations. There are lesser things to contemplate. Maxwell did for electricity and magnetism what Newton had done for the motion of matter. The *Voyager* spacecraft traveling to Jupiter, Saturn, Uranus, and Neptune follows the laws of motion set forth by Newton, and it communicates with Earth by means of the electromagnetic waves predicted by Maxwell.

Maxwell's Equations

Maxwell's equations associated the changes in space and time of electromagnetic phenomena in one uniform theory. The equations, which appeared in Maxwell's *Treatise on Electricity and Magnetism* but were not summarized there, can be stated:

$$\operatorname{curl} \mathbf{H} = \partial D/\partial t + \mathbf{j} \quad (1)$$
$$\operatorname{curl} \mathbf{E} = -\partial \mathbf{B}/\partial t \quad (2)$$
$$\operatorname{div} \mathbf{B} = 0 \quad (3)$$
$$\operatorname{div} \mathbf{D} = \rho \quad (4)$$

H is the magnetic-field strength, **D** is the electric displacement (which is the electric-field strength **E** multiplied by a factor ϵ, whose value depends on the material in which the field exists), t is time, **j** is current density, **B** is the magnetic induction (which is the magnetic-field strength **H** multiplied by a factor μ, whose value depends on the material in which the field exists), and ρ is electric-charge density.

$$\mathbf{D} = \epsilon \mathbf{E} \quad \text{and} \quad \mathbf{B} = \mu \mathbf{H}$$

show the connections between the electric displacement **D** and the electric field **E** and between the magnetic field **H** and the magnetic induction **B**. Vector quantities, which have a magnitude and a direction, are in boldface. Partial differential operations are symbolized as ∂. (A partial derivative is an ordinary derivative taken with respect to one changing variable in an equation with several independent variables; all the variables except the one selected are held constant. Information provided by several partial derivatives can be com- bined so that it is possible to understand how the entire function varies.)

Equation (1) states that any electric current, whether it is a displacement current (a change of electric displacement with time) or a conduction current, is encircled by a magnetic field and that the strength of the magnetic field is determined by the total current it encircles. (The "curl" of an electric or a magnetic field is a differential quantity that takes into account the intensity and the direction of the field.) An electric displacement current is similar in effect to a conduction current such as that produced by electrons moving through a wire, because a magnetic field is produced by either type of current. A displacement current, however, does not involve the movement of

Maxwell's Equations

All science is subject to change as insights deepen. In his special theory of relativity, Einstein taught that Newton's laws are not quite right. In reconciling Maxwell's equations and Newton's laws of motion, he found that it was Newton's laws that had to be modified.

Maxwell's book does not work out all the problems of electric signals and signaling. Rather, it gives us the basis for working them out. That basis is, of course, Maxwell's equations. You will have to look hard to find Maxwell's equations in Maxwell's book; they appear there in pieces, not as we write them today. But they were clear in Maxwell's mind. The equations tell us that a chang-

---

material charges; it can be described as a change in the strength of an electric or a magnetic field.

Similarly, Equation (2) states that a magnetic displacement current, or change in magnetic flux, is encircled by an electric field that is determined by the strength and rate of change of the magnetic field. Equation (3) shows that because free magnetic charges do not exist, the magnetic displacement will pass outward and inward through a closed surface in equal amounts; the displacement does not diverge ("div") from a central charge. (The term *div* also signifies that partial differential operations are performed on the spatial components of **B** and **D**.) Equation (4) states that an electric charge causes an electric field to form in such a way that in every volume of space the charge is accompanied by an electric displacement whose divergence is proportional to the charge.

In a vacuum Equation (4) is zero, because there are no free electric charges. From Equations (1) and (2) it follows that when electric and magnetic fields propagate through space, the electric field is produced by changes in the magnetic field and the magnetic field is produced by changes in the electric field. The electric field and the magnetic field are perpendicular to each other, and both fields are perpendicular to the direction in which the field pattern moves. Maxwell's equations state that the propagation velocity of the fields is finite and that the velocity of propagation is equal to 3×10^8 meters per second, the velocity of light. In addition Maxwell showed that each

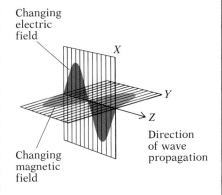

Changing electric field

Changing magnetic field

Direction of wave propagation

field vector obeys a wave equation such that the electric and magnetic fields can vary sinusoidally with time. From these results he concluded that light is an electromagnetic wave.

---

ing electric field will produce a magnetic field, and that a changing magnetic field will produce an electric field. They tell us that patterns of changing electric field and changing magnetic field can travel along together in an endless process as an electromagnetic wave. The speed of electromagnetic waves depends on the material through which they travel, and also on the pattern of electric and magnetic fields. The sort of electromagnetic waves that carry television and radio signals to our homes, or to and from spacecraft, travel at the speed of light (about 3×10^8 meters per second) through space, almost as rapidly through air, and more slowly through transparent solids. Such an electromagnetic wave is slowed by air to 2.99×10^8 meters per second, by water to 2.25×10^8 meters per second, and by quartz to 2×10^8 meters per second.

The electromagnetic waves we consider here are sine waves. Like all sine waves, electromagnetic waves have a wavelength λ, the distance between crests, and a frequency f, the number of crests that pass per second. The wavelength and frequency are related to the velocity (v) by the equation

$$\lambda = \frac{v}{f}$$

The faster the velocity of the wave, the more crests pass per second, and the greater the distance between crests, the fewer crests pass per second.

The Problem of Signaling— According to Maxwell

Maxwell's work tells us why early communicators could not send rapid signals over long distances. In short telegraph lines the cur-

A sine wave has a characteristic wave length, velocity, and frequency. The wavelength is the distance between crests, the velocity v is the speed at which the wave propagates, and the frequency f is the number of crests that pass per second. The relationship between velocity, wavelength, and frequency can be expressed as v = λf, or λ = v/f.

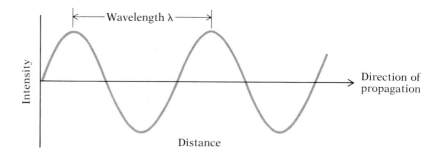

CHAPTER 6

rent seems to be the same all along the line, and the voltage between wire and ground or between wire and wire appears to be the same all along the line. Maxwell's equations do indeed say that this behavior should be expected, but there are two provisos: the electric resistance of the line must be small, and the line must be quite short compared with the wavelength.

The frequencies present in a slowly keyed telegraph signal are very low—not more than 200 Hz, and for a signal of 200 Hz the wavelength is 1.5 million meters, or 934 miles. Thus telegraph lines 100 miles or so long are short compared with the wavelength, but transcontinental or transoceanic circuits are not short compared with the wavelength.

A more serious concern is the resistance of wires. We have seen how Kelvin explained the slowness of signaling along a submarine cable: no appreciable current reaches the far end until enough charge has accumulated along the cable, and this charge accumulates slowly because of the resistance of the wire of the cable. Although the charge will accumulate soon enough in a short line, the delays caused by resistance are serious over long distances.

The early days of undersea cable laying—paying out cable from a three-boat raft.

Maxwell's equations give an explanation of slow signaling that sounds different, though the quantitative prediction is the same. According to his equations, sinusoidal electromagnetic waves have a velocity that may be measured in meters per second, and as they travel, they experience an attenuation that may be measured in deciBels per meter. Both the velocity and the attenuation vary with the frequency of the wave. "Closing the key" momentarily at the transmitting end of a transoceanic cable sends out a pulse of electromagnetic waves, a pulse made up of many frequency components. Because of the resistance of the cable wire, different frequency components travel at different velocities, and they experience different attenuations. Because wave velocity varies with frequency, different frequency components of the transmitted pulse arrive at the receiving end at slightly different times. Much more serious, because attenuation increases with frequency, the higher-frequency components of the transmitted pulse arrive at very weak levels. In early transoceanic cables, high-frequency waves became so weak that the transmitting key had to be closed and opened very slowly to produce signals made up of very low frequency components only.

The Nature of Electromagnetic Waves

Waves are a widespread phenomenon in nature. When we drop a pebble into smooth water, a circle of ripples, or small waves, travels out along the surface. When we speak, sound waves travel outward through the air from our lips. In a similar way, a radio transmitter sends out electromagnetic waves, and the sun sends out light waves, which are also electromagnetic waves. It is our understanding of signals as electromagnetic waves that tells us how to transmit, amplify, guide, and receive them.

A wave is a disturbance of some sort that moves from place to place. A wave in water is not a flow of water; it is a disturbance that travels through the water. The disturbance results from the interaction of two factors: the force of gravity on the piled-up water, which pulls the water down toward its original flat surface, and the momentum or inertia of the moving water, which causes motion to persist in the same way that a bicycle rolls on after we have stopped pedaling. All waves involve at least two such quanti-

As Leonardo da Vinci well understood, waves are not a bodily flow, but a traveling disturbance, like the wind-induced "amber waves of grain." The disturbance pictured here travels outward from a central source as expanding ripples on the surface of still water.

CHAPTER 6

ties. In sound waves they are changes in air pressure due to compression and rarefaction, and the momentum of moving air. As we have noted, the quantities involved in electromagnetic waves are electric and magnetic fields.

Current and voltage are closely related to magnetic and electric fields. A current necessarily produces a magnetic field. Voltage is simply the intensity of the electric field times the distance between two conductors in the direction of the electric field. If we know the electric and magnetic fields around a pair of wires or in a coaxial cable, we can calculate the current flowing along the conductors and the voltage between them.

In discussing electromagnetic waves, we usually talk about waves in which the amplitudes of the electric and magnetic fields vary sinusoidally with time. We can think of this view of electromagnetic waves as following the pattern of our earlier discussion in Chapter 3 of signals in general, in which we resolve the signal into sinusoidal components and ask how the components are modified by circuits as a function of frequency. It is important to point out that in the use of radio and light waves, the bandwidth or range of frequencies needed to represent the voice or TV signal is only a small fraction of the carrier frequency, the mean or nominal frequency at which the transmitter operates. Such electromagnetic waves are almost sinusoidal. The waveform can be described as a sine wave that changes gradually in amplitude and phase. By *gradually* we mean that the change in maximum amplitude or phase from cycle to cycle is very small.

The relation between velocity, wavelength, and frequency ($v = \lambda f$) holds for all sinusoidal waves, including electromagnetic waves. Electric and magnetic fields form patterns that are different for different electromagnetic waves. The various field patterns are known as modes. In plane electromagnetic waves, such as radio and light waves, the electric and magnetic fields are perpendicular to one another, and both are perpendicular to the direction of propagation (the direction in which the wave travels). At any instant the strengths of the electric and magnetic fields do not vary in directions perpendicular to the direction of propagation. The velocity of such plane electromagnetic waves is exactly the velocity of light.

We could argue that there can be no true plane electromagnetic waves of this sort, because their electric and magnetic fields would extend over the breadth and height of the universe. However, far from a transmitter that sends waves out radially the

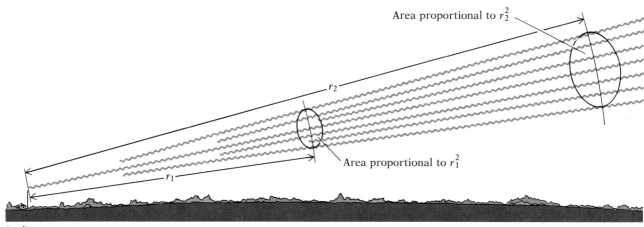

Area proportional to r_2^2

r_2

Area proportional to r_1^2

r_1

Radio
transmitter

Radio waves expand in spheres. At a distance of many wavelengths from a radio transmitter the spheres are so large in size that any small segment of a sphere is almost flat; hence the radio waves propagate along almost parallel lines. If the lines are assumed to be straight, the power of the transmitted radio waves passes through an area that is proportional to the square of the distance r from the transmitter, or r^2, and the power density (power per unit area) is proportional to $1/r^2$. Therefore when a receiving antenna is moved to a distance r_1 away from the transmitter, the power picked up by the antenna varies inversely with distance $1/r_1^2$. Because the power density is proportional to the square of the electric field or the magnetic field generated by the electromagnetic radio signals, the strength of the electric or the magnetic field varies inversely with distance.

waves look and act like plane waves, just as the surface of the spherical earth seems flat.

The Inverse Square Law

As radio waves move outward in all directions from a transmitting antenna, electric and magnetic fields become less intense with distance from the antenna. The product of the electric and magnetic fields, or the square of either, gives us the power density. This measure of power decreases inversely as the square of the distance. This must be so because radio waves expanding outward from a single point flow through successive spheres. The surface areas of the successive spheres increase as the square of the distance, and so the same amount of power is spread out through an ever larger area. The power at a receiving antenna directed at the transmitter is proportional to the power density of the wave reaching that antenna.

Imagine that a receiving antenna a mile away from a radio transmitter receives a power of *P* watts. If we move the antenna to a point 2 miles from the transmitter, it will receive $P/2^2 = P/4$ watts.

The consequences of this inverse-square law can be described in terms of deciBels. A deciBel is 10 times the logarithm of the ratio of power densities (see the table in Chapter 2). If the distance between transmitter and receiver is doubled, the received power goes down to a quarter of its previous value, a decrease of 6 dB. To put this another way, if we double the path length, we must increase the transmitter power by 6 dB to maintain the same received signal power.

The inverse square law accounts for the power required to send signals through space. The *Viking* spacecraft could send signals back from Mars with a power of only 20 watts. The more sophisticated *Voyager* sent signals back from more distant Jupiter also with a power of 20 watts. *Voyager* sent signals back from Saturn, signals 14 dB weaker at Earth, and from Uranus, signals 23 dB weaker at Earth, and from Neptune, signals 30 dB weaker at Earth. Somewhat better antennas and receivers were used than had been to receive the signals from *Voyager* at Jupiter. Chiefly, the signals had to be sent back at a slower rate.

We are not as lucky in radio communication between points on or near the surface of the earth, because intervening hills and, indeed, the bulge of the earth cast shadows.

Diffraction and Shadows

Today we know that light consists of electromagnetic waves. But the fact that light casts shadows caused the great Newton himself to reject the idea that light might be a wave of some sort. Instead, he believed that light was a flow of tiny particles that he called corpuscles. A part of wave theory called diffraction theory, which applies to electromagnetic waves and all other waves, tells us that the shadows cast by waves are a little fuzzy. However, when the wavelength is very short compared with the size of an object casting a shadow, and compared to the distance from the object to the shadow, the shadow is sharp. The wavelength of light is very short, and so the shadow of your hand on a surface a foot away is very sharp if the source of light is very small. Radio waves are much longer, and the shadows they cast are fuzzier. Very long radio waves creep around hills and buildings, so to speak, and we can receive them on the far side.

Because of shadows and diffraction, the inverse square law holds only if the transmitter is visible from the receiver. Intervening hills or the bulge of the earth can weaken the signal drastically. As radio waves are made shorter and shorter, they behave very much like light. Very short radio waves cast sharp shadows.

Microwave antennas must be in sight of each other in order to send and receive narrow beams of microwave signals, so on Earth they are usually mounted on tall buildings or towers, such as this parabolic reflector over London. In space, microwave antennas are placed on orbiting communication satellites.

Directive Radio Antennas

A directive transmitting antenna sends out radio waves in a horizontal plane or in a narrow beam rather than in all directions. Waves of short wavelength are often completely blocked by obstructions in the transmission path, in the same way that shadows are created when light is obstructed. By permitting a larger fraction of the transmitted power to be picked up by the receiving antenna, directive antennas compensate for interference.

At 100 MHz, the frequency of frequency modulated (FM) signals and very high frequency (VHF) television signals, a wavelength is about three meters; the broadcast antennas are roughly cylindrical in configuration, producing waves that spread equally in all directions. To maximize the fraction of waves received from the transmitting station FM and VHF television receiving antennas are pointed in the direction of the transmitter, and tapered as shown in drawing A.

Good reception of ultra high frequency (UHF) television signals is more difficult to achieve. With an average wavelength of only 0.3 meter and a frequency of 500 MHz, UHF signals are significantly weakened by intervening hills or structures. Usually UHF rooftop receiving antennas are incorporated with VHF antennas in a complicated way. A "settop" UHF antenna is commonly a single loop of wire about 19 centimeters in diameter (drawing B).

The microwaves that are used in the long-distance relay of telephone and television signals have a wavelength of only 0.08 meter and so are easily obstructed. The transmitting antenna and the receiving antenna for such a system are highly directive and are placed on hilltops 20 to 30 miles apart. Microwave transmitting antennas sometimes consist of a metal horn (drawing C), which, like a megaphone, guides waves in one direction. The horn feeds the waves into a concave dish reflector; the reflector focuses the waves into a narrow beam, which is directed toward the receiver. The receiving antenna reverses the transmitting proce-

The VHF (very high frequency) television signals, which have a wavelength of about 3 meters, reach home television antennas even when they are somewhat shadowed by hills. UHF (ultra high frequency) television signals, which have a wavelength of about 0.3 meters, have sharper shadows, and a television set has to be almost in line of sight of the transmitter in order to get a good signal. Because microwaves have an even shorter wavelength of 0.03 to 0.3 meters, they can be received only if the transmitter can be seen clearly from the receiver. In microwave radio systems that carry telephone and television signals across the country, successive receivers, amplifiers, and transmitters must be placed on hilltops 20 to 30 miles apart, so that each receiving antenna is in clear view of a transmitting antenna.

dure to recover the signal. The same antenna is used to transmit at one frequency and to receive at another. In general the wavelength λ of the signal and the diameter D of the transmitting antenna determine the width W of the microwave beam at the receiving unit a distance L away, as expressed by the following equation:

$$W = \frac{\lambda L}{D}$$

The relation between the power picked up by the receiving antenna (P_R) and the power sent out by the transmitting antenna (P_T) is a function of the wavelength λ, the size of the antennas, and the distance L

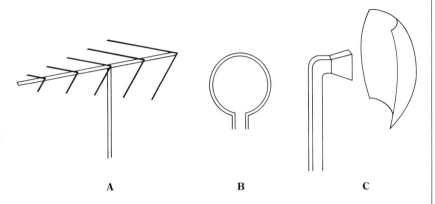

A   **B**   **C**

between them. A_T and A_R represent the effective areas of the transmitting antenna and the receiving antenna respectively.

$$\frac{P_R}{P_T} = \frac{A_T A_R}{\lambda^2 L^2}$$

This simple and useful transmission formula for line-of-sight antennas was first given by Harald T. Friis of Bell Laboratories.

Antennas

On earth or in space, the distance over which we can send a radio signal depends on more than transmitter power. It depends on the bandwidth and signal-to-noise ratio required. It also depends on the configuration of the transmitting and receiving antennas.

We will receive a larger fraction of the transmitted power if we use a directive transmitting antenna and point it at the receiving antenna. A directive antenna is one that sends out radio waves in a narrow beam rather than in all directions. The power density in the beam of a directive antenna falls off inversely with the square of the distance, and so does the power received by a receiving antenna, but the power density and the power received are greater at any distance for a directive transmitting antenna than for one that sends power out in all directions.

For an antenna to be directive, it must be larger than a wavelength, and the larger it is the more directive it can be. At comparatively long wavelengths such as those used in AM radio broadcast, transmitting antennas are tall towers that send signals almost equally toward all points of the compass. A good deal of power goes upward. The more complicated VHF and UHF television transmitting antennas mainly send power out horizontally, often more in some chosen direction than in others. Microwave antennas send out narrow radio beams by focusing the signals with concave metal reflectors.

An antenna that receives a narrow beam of radio waves picks up more power when it is pointed at the radio transmitter. Receiving antennas rely on concave reflectors to capture a signal. As we might expect, the amount of power picked up is proportional to the area of the reflector; it is very close to the area times the power density of the radio wave reaching the antenna. Thus, microwave receiving antennas have large concave reflectors, larger for radio astronomy, communication with spacecraft, and receiving signals via satellites than for point-to-point communication on earth.

Satellites are ideal locations for microwave repeaters. A receiver on a satellite can pick up a signal transmitted from one country, feed it to an amplifier, and send the amplified signal to another country. The satellite is in a clear line of sight of both countries. Moreover, signals from a satellite go through the earth's atmosphere at a fairly steep angle, and thereby avoid a phenomenon that plagues earthbound microwave transmission. Terrestrial microwave signals suffer from fading. Layers of different densities in the atmosphere can so bend the paths of radio waves that signals from the same source arrive from two directions along paths of slightly different lengths, and the two signals may be out of phase and cancel one another. Over desert flats a wave reflected from the surface of the earth may cancel a radio wave going directly from antenna to antenna.

Guided Waves

We have seen that radio waves are electromagnetic waves that travel out through space from their source. Electromagnetic waves can also be guided by wires, tubes, or transparent fibers. Guided transmission provides most telephone service and thus most of our electronic personal communication. Through cable TV, guided transmission also plays an increasing part in our few-to-many mass communication.

In technologically advanced areas of the world, the future belongs to digital communication by electromagnetic waves guided by highly transparent optical fibers. It seems sensible, however, to discuss the guiding of electromagnetic waves as it unfolded through time. Let us, then, first consider guided transmission of electromagnetic waves by means of a wire over a sheet of metal; this approximates early telegraph wires that were strung over the conducting earth. A similar arrangement is found in today's integrated circuits, which contain tiny conducting strips on a dielectric (nonconducting) layer over a conducting sheet.

The waves guided by a wire over a conducting sheet are plane electromagnetic waves, just as light and radio waves are. As we have noted, the electric and magnetic fields are perpendicular to each other, and they are everywhere normal (perpendicular to, at right angles to) the direction of propagation, which is the direction in which the wave travels, or the direction of the wire. The electric and magnetic fields become weaker and weaker farther and farther away from the wire.

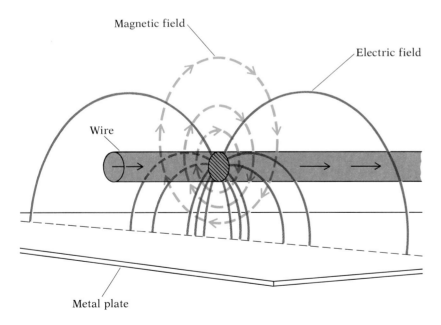

Magnetic field

Electric field

Wire

Metal plate

Electromagnetic waves can be guided by a wire placed over a metal plate. The magnetic field of the wave traveling along the wire closes in loops around the wire (green arrows), and a current flows in the wire, so that there is no magnetic field inside. The electric field (purple lines) starts at the electric charges on the wire and terminates at electric charges on the metal plate. The electric and magnetic fields are everywhere perpendicular to one another (a transverse electromagnetic wave), and they weaken with distance from the wire.

Plane electromagnetic waves can also travel over a pair of wires, the configuration that in telephony replaced the wire over ground. The field pattern is basically the same as that for a wire over a metal sheet, if we regard the field pattern as duplicated on the lower side of a plane placed between the two wires.

A plane electromagnetic wave can also be guided by a coaxial, a metal tube with a wire running down the center, as we noted in Chapter 5.

If a wire over a conducting sheet, or a pair of wires, or a coaxial could be made of perfectly conducting material, waves of all frequencies would travel along them with exactly the velocity of light. Actual transmission lines are made of copper, which is not a

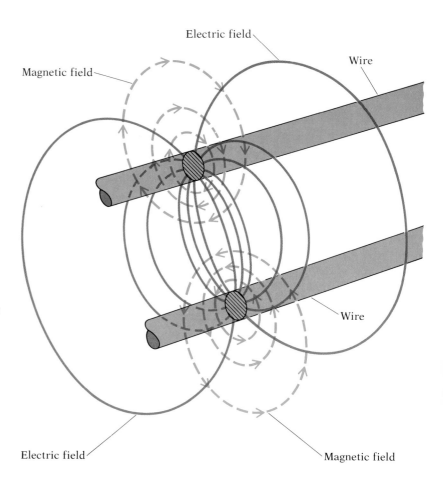

An electromagnetic wave can travel guided by two wires. The electric and magnetic fields have the same shapes as in the case of a wire above a metal surface. Think of the electric field as reflected in the metal surface so that its direction is the same below the surface as above it. The current in the upper wire is reflected in the lower wire as a current in the opposite direction. The magnetic field below the reflector has the same shape but the opposite direction as that above the reflector.

CHAPTER 6

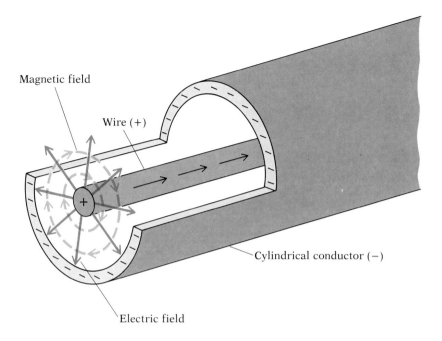

Magnetic field

Wire (+)

Cylindrical conductor (−)

Electric field

A *coaxial cable consists of a central conducting wire separated from an outer conducting cylinder by an insulator. The central conductor is positive with respect to the outer conductor and carries a current (black arrows). An electromagnetic wave guided by the cable travels in the space between the two conductors. When a wave travels along the cable, electric and magnetic fields form between the wire and the cylinder. The electric field is radial and is directed outward (purple arrows); the magnetic field is circumferential and is directed clockwise (green arrows).*

perfect conductor, and so a wave loses energy as it travels along a pair of wires or a coaxial. The wave loses power, and its amplitude decreases. The loss increases with frequency. In a typical coaxial, for example, the loss is about 33 dB per mile (a loss of power by a factor of 2000) at a frequency of 70 MHz. The loss in deciBels increases with the square root of the frequency.

Delay Distortion

In an imperfectly conducting cable, sinusoidal waves of different frequencies travel with slightly different velocities, and this phenomenon gives rise to something called delay distortion. To illustrate it, suppose that we send a short pulse into a transmission line. The pulse can be represented by many sinusoidal components with different frequencies. The sinusoidal components will travel along the transmission line with slightly different velocities. Some components will reach the far end later than others, so that all the components will not peak at the same time. The received pulse will be longer or broader than the transmitted pulse. In television such delay distortion will cause a blurring of the image.

Submarine repeaters ready for laying.

The received pulse is broader for another reason. Higher-frequency components are attenuated more than lower-frequency components, and they arrive too weak to reproduce the sharp peak of the transmitted pulse.

In some degree, variations of velocity and frequency in transmission can be corrected by networks or circuits called equalizers. But because attenuation increases with frequency, there is a practical limit on the band of frequencies that can be transmitted. This limit was pushed upward by putting repeaters closer and closer together—the repeater spacing in the L5 coaxial cable system is one mile. But it is awkward to have to amplify a signal so frequently in sending it thousands of miles.

Microwave Waveguides

Some way was needed to achieve broader bandwidths so that we could send telephone *and* television signals over long distance economically. The answer would be circuits with less attenuation at broad bandwidths. Engineers investigated sending microwave signals through rectangular or circular conducting tubes called waveguides. Ever since World War II, such waveguides have been used in microwave radar and radio to connect antennas to microwave transmitters and receivers.

Only very high frequency waves can go through a metal tube of reasonable size, for below a certain cutoff frequency propagation is impossible.

The existence of a cutoff frequency, which is inversely proportional to a transverse dimension, is characteristic of all waveguides, regardless of the shape of the cross section. So is the variation of velocity with frequency.

The propagation of electromagnetic waves through a waveguide has another peculiarity. If the frequency is high enough, different field patterns or modes can travel through the same waveguide, although at different velocities. When there are many modes of propagation, even if the initial signal is sent out in just one mode, irregularities in the waveguide can put part of the signal energy into other modes with other velocities.

At many laboratories ingenious people worked for many years toward such a high-capacity, long-distance waveguide system. An experimental system was built. But something better came along, and waveguide is an art of the past—for long-haul communication systems, at any rate.

Optical Fibers

Almost the whole future of communications revolves around the new technology that usurped waveguides. That technology is communication by means of light waves traveling through transparent optical fibers.

Communication by means of light waves is not new. In the days of the ancient Silk Road linking China and the West, news of attacks was sent by beacon fires. We have all heard of semaphore telegraphs, signal flags, and heliograph mirrors flashing in the sun. With his photophone, Bell experimented with the transmission of voice signals by means of reflected sunlight. The light of the sun and stars travels far through empty space, and on clear days light can travel far through clear air. But, the idea of sensing light flashes many miles away through a thread of glass might well stagger the imagination.

The idea of guiding light is not all that new. In 1881 William Wheeler, a Concord, Massachusetts engineer, patented a system of internally reflecting pipes, which would distribute light within a building in the days before Edison's electric bulb. That same year Charles Vernon Boys, a British physicist, described the transmission of light through thin fibers of glass. But no one thought of this as a means of communicating over long distances. Too much light was absorbed in traveling through the fiber for *that*.

In 1966 Charles Kuen Kao and G. A. Hockman of Standard Telephones, Ltd. thought about this matter rather than easily accepting such an assumption. They looked into known mechanisms of attenuation or energy loss in the transmission of light through very pure quartz glass. They pointed out that in principle the attenuation could be very low, and predicted an attenuation as low as 20 dB/km, far less than had ever been demonstrated. Many people read, believed, and set to work.

Surprisingly to some, far lower attenuations have been attained. By 1968 researchers had prepared samples of bulk silica (quartz) with an attenuation of 5 dB/km. In 1970 Corning Glass Works prepared fibers with 20 dB/km or less attenuation. In the same year, I. Hayashi and others at Bell Laboratories demonstrated a semiconductor injection laser, a suitable low-voltage light source. Today, optical fibers have attenuations as low as 0.16 dB/km. Lower losses are predicted for other glasses. Reliable semiconductor lasers have been developed that operate at wavelengths best suited for long-haul communication.

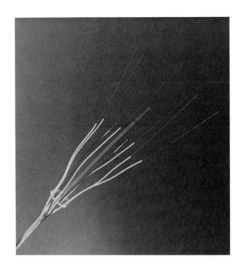

Optical fiber cable consisting of six optical fibers sheathed in protective coatings, copper wires to supply electric power to repeaters, and a central strong wire.

As improvements in the manufacture of glass culminated in the creation of optical fibers, the loss of light intensity with distance has decreased rapidly over the years. Today about 96 percent of the light passes through a kilometer length of fiber.

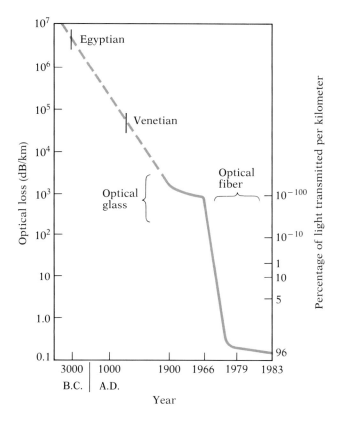

The optical fibers used in communication consist of three concentric regions: the core, the cladding (both of glass), and a jacket of organic material. The outer jacket gives mechanical strength and protection and ensures that the fiber is optically isolated from other fibers. The core and cladding are of very pure silica glass, with a few carefully chosen and controlled dopants added to alter the index of refraction. The index of refraction tells how much a ray of light will bend when entering a substance obliquely from air. It is the ratio of the velocity of light in a vacuum to the velocity of light in the substance. The core has a higher index of refraction than the cladding; this is necessary for the fiber to guide light. The electric and magnetic fields in the core are very much like those of a plane electromagnetic wave in space—with the addition of some small electric and magnetic field components parallel to the axis. In the lower-index region outside of the core the fields fall off rapidly with radial distance.

CHAPTER 6

In fibers used for transmission over long distances, the central core is extremely small. A core of such smallness ensures that there is only one mode of transmission. Pulses will not be broadened by the excitation of various modes with various velocities.

Fibers whose cores have a diameter many times the wavelength of light can successfully transmit light waves over short or moderate distances. For such fibers we can if we wish describe the guided signal as a summation of various modes, each with a particular field pattern and velocity. However, there is another intuitive and attractive way of thinking of the travel of light through large-core fibers—that is, in terms of light rays.

Think of a core of constant index of refraction and surrounding glass of a lower index of refraction, with a sharp boundary between them. In such a stepped-index fiber, light rays propagate through the core in straight lines. Light rays reaching the boundary of the cladding at a small angle will be totally reflected back into the core.

Total reflection takes place when rays of light in any medium of higher index of refraction reach a boundary with a medium of lower index of refraction at less than a *critical angle*. This was discovered by the British physicist John Tyndall in 1870 for light traveling in water at an air-water boundary. It is experienced by divers who look toward the surface of still water above at an angle close to the horizontal. The diver can't see through the surface.

In some large-core multimode fibers, the index of refraction decreases smoothly from the center to the cladding; thereafter, the index of refraction is constant. The core of such a graded-index fiber acts as a long, continuous converging lens, which urges toward the axis light rays that would otherwise diverge. Travel time varies less with ray angle in the graded-index fiber than in the stepped-index fiber.

It is, however, single-mode fibers that make economical long-distance communication possible. Typically (as in a fiber-optic submarine cable) the operating wavelength will be 1.3 micrometers (millionths of a meter—a micrometer is 40 millionths of an inch), and the diameter of the inner core of the fiber about 8 micrometers in diameter. Although the rate of transmission is theoretically limited to around 200 gigabits a second (200,000 million pulses per second), present-day practical systems can transmit up to around 2 gigabits a second. At this rate, one cable can carry 30,000 telephone channels. The number of channels will increase with time.

Single mode

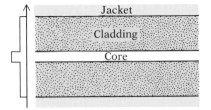

Multimode—stepped index

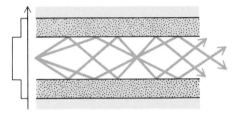

Multimode—graded index

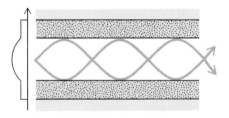

Optical fibers consist of a small central core of higher index of refraction surrounded by a region of lower index of refraction called the cladding. In the single-mode fiber, light travels through the core parallel to the axis. In the stepped-index multimode fiber, light rays propagate through the core in straight lines, reflecting off the sharp boundary between the core and the lower-index cladding. In the graded-index multimode fiber, the core acts as a very long converging lens continually urging light rays toward the center. The light rays actually travel at very slight angles, much smaller than shown here.

Submarine Cable Systems Under the Atlantic

SYSTEM NAME	FIRST USED	TERMINALS	VOICE CHANNEL CAPACITY		REPEATER SPACING (STATUTE MILES)	TECHNOLOGY
			SYSTEM DESIGN	MAXIMUM TASI		
TAT-1*	1958	Scotland to Newfoundland	36	72	44	2 separate coaxials
TAT-2*	1959	France to Newfoundland	36	72	44	2 separate coaxials
TAT-3*	1963	United Kingdom to New Jersey	138	276	23	one coaxial
TAT-4*	1965	France to New Jersey	138	345	23	one coaxial
TAT-5	1970	Spain to Rhode Island	845	2,112	12	one coaxial
TAT-6	1976	France to Rhode Island	4,000	10,000	6	one coaxial
TAT-7	1983	United Kingdom to New Jersey	4,200	10,500	6	one coaxial
TAT-8	1988	United Kingdom and France to New Jersey		40,000	41	fiber

*Retired.

The technology of fiber optics has advanced rapidly. The accomplishments of the research laboratory move into the market in a few years. The glass of the fiber is so pure and so nearly lossless that light can be transmitted and detected over fiber lengths of hundreds of miles. The capacity of fiber can be increased by using simultaneously on the same fiber two or more lasers operating at slightly different wavelengths. Where can this lead?

Today undersea optical fiber cables link continents. Hundreds of thousands of miles of optical fiber cable have been laid in the United States, Europe, and Japan. Most long-distance telephone calls travel over optical fibers. Optical fiber cables carry much traffic within and between telephone offices. In general, undersea systems provide fewer channels than overland systems, whether the systems use fiber optics or coaxial cables. Optical fibers are particularly advantageous for underseas transmission because they can attain a high capacity yet they transmit well with a large repeater spacing. TAT-8 is a digital system using optical fiber and solid-state lasers. Two pairs of fiber are in active use, with an additional pair as protection. A new underseas fiber system, TAT-9, is planned for use in 1991. It will offer 80,000 voice channels with repeaters spaced every 75 statute miles. Undersea cable systems employ Time Assignment Speech Interpolation (TASI) to increase the number of voice channels by using the silent intervals in normal speech to carry active portions of other conversations.

Fiber transmission to homes has been tried experimentally. It may be the wave of the future, but subscriber loops made of twisted wire pairs already reach most homes. Besides carrying voice calls, twisted pairs carry facsimile and computer mail at rates up to millions of bits a second. How are we to justify economically replacing twisted pairs with optical fibers?

We can't predict the future in detail, but we can be reasonably sure that transmission by electromagnetic waves is here to stay. But, efforts to communicate would be in vain if we could not generate, receive, amplify, and manipulate those waves. In the next chapter we shall discuss the devices that make those processes possible.

British Telecom's cableship Alert *laying the first international optical fiber cable between Broadstairs in Kent and Ostend, Belgium.*

7

Electric and Electronic Devices: The Physics of Communication Systems

The transmission of signals by radio, by wire, or by transparent fiber is a phenomenon based on and described by the laws of physics. The devices used to generate, amplify, and detect signals are also based on the laws of physics. All the services that communication performs for us exist only because we have discovered and understood various physical phenomena and have harnessed them to our use in ingenious ways. The variety of physical phenomena that we exploit has increased over the years, as have the complexity and capability of the devices that make use of them. In the earliest days of electric communication the devices were very primitive. Modern devices are, by comparison, very sophisticated. Yet, all behave, as they must, in accord with the laws of physics, which are a part of the laws of nature.

An early triode vacuum tube. Over a 40-year period, vacuum tubes were our essential resource in realizing long-distance and multiplex telephony, television, and fast electronic computers.

Hertz and Electromagnetic Waves

In 1864 Maxwell read at a meeting of the Royal Society a paper on his dynamical theory of the electromagnetic field; the paper was published in 1865. Maxwell predicted a new sort of wave, an electromagnetic wave made up of rapidly changing electric and magnetic fields traveling with the speed of light. According to Maxwell, as an electromagnetic wave travels, the changing electric field produces the magnetic field, and the changing magnetic field produces the electric field. Scientists were interested but unconvinced. From 1886 to 1891, the German physicist Heinrich Hertz undertook, at the University of Karlsruhe, a sequence of investigations that proved the existence of electromagnetic waves and showed that they could be reflected, refracted, and focused much as light waves are.

Hertz reasoned that electromagnetic waves should be produced when a current oscillates back and forth in a wire or loop of wire. The oscillating current creates a changing magnetic field, and the changing magnetic field creates a changing electric field. The changing electric and magnetic fields should produce an electromagnetic wave. Electromagnetic waves should be radiated (and collected) most efficiently when the wire or loop is resonant, so that a current, once started, surges back and forth naturally, as a pendulum swings back and forth.

The device that Hertz used to generate electromagnetic waves was the spark-coil transmitter. Hertz divided a wire in half, leaving a gap in the center. He used a spark coil to build up a high voltage between the two halves of the wire. Positive charge accumulated on one half of the wire, and negative charge on the other half. The sections of wire were now acting as a capacitor. When the voltage became high enough, a spark jumped across the gap. The two wires were in effect connected as one wire, and the stored charge flowed periodically back and forth as an electric current in the wire. This oscillating electric current produced an oscillating magnetic field in the wire, and caused the radiation of electromagnetic waves. The frequency of oscillation decided the wavelength of the electromagnetic radiation, which turned out to be twice the total length of the wire. Eventually, the oscillation grew weaker, the spark stopped, and the voltage between the two wires started to build up again. In a modified form, that means of generating radio waves survived through World War I as the spark transmitter.

Hertz detected the electromagnetic waves by means of a very

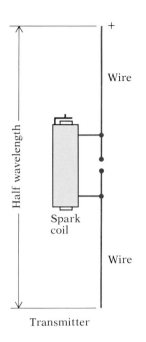

Transmitter

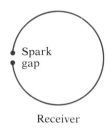

Receiver

simple receiver: a circular loop of heavy wire interrupted by a very narrow gap. The waves sent out by the spark transmitter caused an oscillation in the loop, and a visible spark jumped across the tiny gap in the loop of wire.

Hertz's primitive equipment enabled him to perform crucial experiments, which showed that the waves he produced conformed to Maxwell's equations and did indeed behave just as light would—if light had a far greater wavelength. But Hertz's apparatus was of use only in laboratory experiments because it produced electromagnetic waves that were too feeble and of too short a wavelength to be of use in early wireless communication.

Marconi and the Wireless

Guglielmo Marconi's contribution to communication was the establishment of wireless telegraphy. He gathered together the ideas and devices of others, then added his own to invent the first instrument that communicated by sending electromagnetic waves through the air. His demonstrations of wireless telegraphy proved conclusively and profitably that it could be of use to man.

Guglielmo Marconi

Although Maxwell discovered the equations that describe electromagnetic waves and Hertz first produced electromagnetic waves, Guglielmo Marconi was the first to find a practical use for them when he invented the wireless telegraph, which came to be known as the radio. Marconi was born in Bologna, Italy, in 1874. He had no formal university education. When he was 20 years old, he set himself the task of inventing a device that utilized electromagnetic waves for communication. He worked frantically, feeling that others must be trying to do the same thing. No one was. By 1895 he had added a ground and a tall antenna to Hertz's original spark transmitter design and with this device had transmitted radio waves over a distance of more than a kilometer. Because interest in his invention was lacking in Italy, Marconi went to England; there in 1896 he obtained the first patent on the wireless. By 1897 he had transmitted

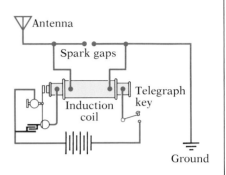

messages over a distance of close to 20 kilometers, and in that same year Marconi's Wireless Telegraph Co., Ltd., was formed. Successful communication with ships beyond the horizon encouraged Marconi in 1900 to undertake signaling across the Atlantic. He succeeded in 1901. In the photograph, Marconi is seated at the receiving set in St. John's, Newfoundland. In 1909 he received jointly with C. F. Braun the Nobel prize in physics. Marconi was made a marchese and a member of the Italian senate, and he served as president of the Royal Italian Academy. He died in Rome on July 20, 1937.

Marconi used Hertz's spark oscillator as a transmitter. The operator generated signals by stopping and starting spark discharges with a telegraph key. By this means, he could produce long and short bursts of electromagnetic waves that corresponded to the dots and dashes of Morse code. To this transmitter, Marconi added something important. He devised a tall wire antenna, which he connected to one side of the spark gap; he connected the other side to the earth, which acted as a second wire. The electric oscillations in the tall antenna had a lower frequency and a longer wavelength than those that Hertz had studied, and these radio waves were more powerful and could be detected farther away. Happily, Marconi had in effect made the antenna twice as long as the wire; the conducting earth acted as a reflector or mirror, and formed an electrical image of the antenna above it, so that electromagnetic waves were radiated as if from two wires of equal length connected by the transmitter in the center.

Among the devices that Maconi gathered were better instruments than Hertz had used for detecting electromagnetic waves. In 1890 Edouard Branly, a French physicist, invented a far more sensitive means for detecting radio waves: the coherer. In a coherer radio waves make a pinch of metal filings stick together so that the filings form a conducting path. In the absence of a radio wave, the filings touch one another so lightly that they cannot conduct electricity appreciably. When stuck together, the metal filings complete a circuit that carries the arriving signal to a pair of headphones.

In turn, the coherer was eventually replaced by the crystal detector, the first semiconducting solid-state device used in radio. Initially, the physics of the operation of the crystal detector was not understood. Its function was; the crystal detector is a sensitive *rectifier* that allows electric current to pass in one direction only. The crystal detector turned a radio-frequency dot or dash from the spark transmitter into a noisy unidirectional current that could be heard through a pair of headphones.

In 1879 Sir Oliver Joseph Lodge, a British physicist who had improved Branly's coherer, patented the use of resonant circuits consisting of inductors and capacitors to set the frequency of oscillation accurately and to tune a receiver to that frequency. Marconi soon used tuning effectively, and disputed its origin. In a sense tuning was not a new idea, for Hertz's receiving circuits had been tuned by their very nature, and Hertz had investigated the effects of tuning on a received signal by altering the size of the receiving circuit. The deliberate use of tuning made it possible for transmitters to send signals on different frequencies so that several transmitters could function simultaneously without interfering with one another. Circuits similar to those of Lodge and Marconi are used today.

Marconi's crystal detector, the first semiconductor device used in radio. It changed the incoherent radio waves received from the spark transmitter into noisy dots and dashes.

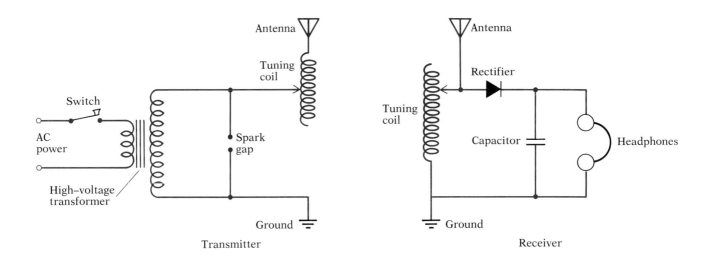

Transmitter

Receiver

In the early radio transmitter (left), a high voltage between two wires produced a spark as in Hertz's transmitter, but the two wires separated by the spark gap were an antenna and the ground. A spark jumping across the spark gap connected the antenna to the ground through a tuning coil, an inductor that could be adjusted to change the radio frequency of the transmitter. Electromagnetic waves radiated from the gap and excited a current in a resonant receiving circuit (right). The voltage across the receiver tuning coil caused a current to flow through the rectifier from a sharply pointed wire, or "cat whisker." Each burst of oscillation from the transmitter caused a rectified current to flow through the headphones. Some of the current was stored in the large capacitor as a charge until it could pass through the headphones of the radio receiver.

Continuous Waves

Spark oscillators produce a succession of bursts of radio-frequency power. The electric oscillations of successive bursts are incoherent; that is, the bursts of oscillating current following successive sparks have random phases with respect to one another, as do the radiating electromagnetic waves. When we pick up an electromagnetic wave with an antenna, receive it as an oscillating current, and then rectify it and listen to it with headphones, dots and dashes from a spark transmitter sound like short or long bursts of noise. The noisy signal produced by a spark transmitter cannot be used for telephony.

The limitations of the spark oscillator were recognized, but no one knew how to produce a steady, coherent sinusoidal radio wave until 1903. In that year Valdemar Poulsen, a Danish engineer, devised his oscillating arc. (Poulsen later invented the wire recorder, the forerunner of today's tape recorder.) When supplied by direct current, Poulsen's oscillating arc produced continuous radio waves. Poulsen's arc was used in radio telegraphy, and in early experiments in sending voice by radio.

In 1906 Ernst Frederik Werner Alexanderson of the General Electric Company demonstrated a new way to produce continuous radio waves. He built an alternator, or rotating alternating-current generator, for radio transmission. It produced an alternating current that changed its direction of flow smoothly and sinu-

soidally with time. His alternator may be likened to alternators used in power stations, which supply a current of 50 Hz or 60 Hz to our homes. Alexanderson's alternator, however, produced radio waves with a frequency of 50,000 Hz at one kilowatt of power. In the same year Reginald Aubrey Fessenden, a radio pioneer who was the most active proponent of continuous waves in the United States, used the alternator to transmit speech and music.

Alexanderson built several other alternators, gradually increasing both power and frequency. Waves of higher power could reach larger distances. In 1915 he built a 50-kilowatt, 50-kilohertz alternator, and in 1918 a 200-kilowatt alternator. His 200-kilowatt alternators operated at frequencies of 12.5 or 28.57 kilohertz, the same frequencies now used for navigational radio but only about a twentieth as high as the low end of the broadcast band. No one before him had been able to make workable alternators of such high frequency.

Alexanderson's alternators were used during the latter part of World War I in communicating with Germany about ending the war. The alternators were operated until 1948. By that time a remarkable invention had displaced all other radio transmitters. That invention was, of course, the triode vacuum tube. It provided the solution to the problems of generating continuous radio waves and of amplifying and detecting radio signals. Indeed, the vacuum tube solved the problem of producing, amplifying, and processing all sorts of signals.

De Forest and the Vacuum Tube

Lee de Forest invented the vacuum tube in 1907. Until the invention of the transistor after World War II, all of radio, long-distance telephony, and complicated electronics, including electronic computers, derived from de Forest's invention. It is one of the very greats in the history of technology.

The vacuum tube, or audion as de Forest called it, resembled superficially an earlier device made by John Ambrose Fleming in England, although Fleming's device could not amplify signals and

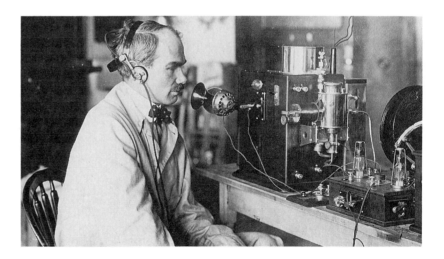

Lee de Forest, inventor of the vacuum tube, with an early radio transmitter and receiver.

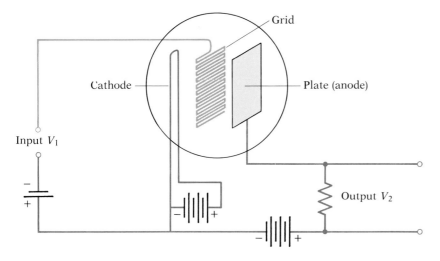

Grid

Cathode

Plate (anode)

Input V_1

Output V_2

In a simple vacuum tube, electrons emitted from the negatively charged cathode are attracted to the positively charged anode. A grid placed between the cathode and plate controls the flow of electrons. In the absence of an incoming signal, the highly negative grid prevents many electrons from flowing to the cathode. When a positive signal of voltage V_1 is applied to the grid, the grid becomes less negatively charged with respect to the cathode, and a larger number of electrons flow from cathode to plate and produce a larger change in voltage across resistor R. The signal is amplified because the change in voltage output V_2 is greater than the change in input voltage V_1.

de Forest's could. The resemblance between the two inventions probably explains why de Forest did not receive the Nobel prize. In 1913 Harry De Forest Arnold (unrelated to Lee de Forest) of AT&T and Irving Langmuir of General Electric independently produced improved vacuum tubes with a hard vacuum (containing almost no gas).

A simple vacuum tube, or triode, consists of three essential elements: a hot cathode, a grid of parallel wires, and a plate, or anode. The hot cathode emits electrons, which form the current. The grid of parallel wires controls the flow of electrons, and the plate collects the electrons. The cathode, grid, and plate are all sealed in an evacuated tube made of glass or metal, and wires sealed through the glass connect the cathode, grid, and plate to external batteries and circuits. In the simple vacuum tube shown on this page, the cathode is a metal filament, like that of a light bulb, heated by current from a battery. The plate is held positive with respect to the cathode by a second battery or voltage source. The plate attracts electrons from the cathode. The grid is held negative with respect to the cathode by a battery or another voltage source. It tends to repel electrons, although if the voltage to the grid is low, electrons will be able to pass through it to the cathode. The grid is placed close to the cathode so that a small change in voltage of the grid has a large effect on the number of electrons that can cross to the plate: the more negative the grid, the less the current that flows to the plate; the less negative the grid, the greater the current that flows to the plate. Incoming signals go directly to the grid. When a positive input signal voltage V_1 is applied to the grid, it becomes less negatively charged with respect to the cathode, and a larger number of electrons flow from cathode to plate. This causes a change in voltage across the resistor. The change in output voltage V_2 is greater than the change in input voltage V_1 and so the tube acts as an amplifier.

The vacuum tube can do all sorts of wonderful things. In a radio receiver it can amplify the weak radio-frequency signal picked up by the antenna. A vacuum tube can rectify a radio-frequency voice signal or a television signal and recover the corresponding baseband signal. Another vacuum tube can then amplify the recovered baseband signal. The output of a vacuum-tube amplifier can be connected to the input by a resonant circuit in such a way that the feedback strengthens the output. The electric circuit formed by the connection produces an oscillating current at a fre-

A high-power klystron used for ultrahigh frequency television transmission.

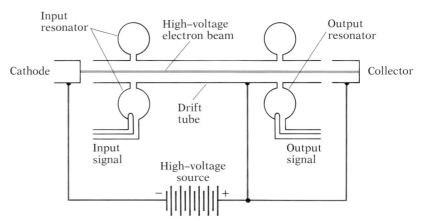

The klystron is an extremely high power vacuum tube that will amplify microwaves. A high-voltage beam of electrons from a cathode is passed through an input resonator, where the electrons are alternately speeded up and slowed down by a sinusoidal electric field that varies at a frequency equal to that of a microwave input signal. As the electrons pass through a drift tube on their way to a collector electrode, the speeded-up electrons overtake the slower electrons so that electrons gather in bunches, forming an alternating current at the frequency of the input signal. When the alternating current flows across the gap of an output resonator, it sets up a strong electric field inside the resonator. This field is stronger than the original electric field, and so the output signal, which is tapped from the surplus energy of the output electric field, is greatly amplified.

quency determined by the resonant circuit. Such a vacuum-tube oscillator can be used as a source of radio waves. For example, it can generate the sinusoidal wave used to change the frequency of signals, as described in Chapter 5 in connection with multiplexing.

The performance of early vacuum tubes was poor. They would not operate at very high radio frequencies, and their gain as amplifiers was rather low. That is, the ratio of output power to input power was small. The power that could be produced was small. Vacuum tubes evolved in two directions. One was toward higher gain and broader bandwidth at low power levels, attained in the tetrode and pentode through additional grids. The other direction was toward higher power.

The vacuum tubes that trace their descent directly from de Forest's audion are obsolete except for high-power tubes used in radio transmitters. Transistors have taken their place. Other forms of vacuum tubes still in use are klystrons and traveling-wave tubes, which operate at microwave (super-high) frequencies. Klystrons produce the megawatts of microwave power needed for radar, space communication, and linear accelerators. Traveling-wave tubes give a high gain over a broad band of frequencies. They are used to produce around 10 watts of power for transmission from communication satellites and space vehicles. The magnetron oscillators that made World War II microwave radar possible have been relegated to microwave ovens.

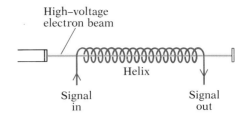

The traveling-wave tube is a type of vacuum tube that gives high gain over a broad band of frequencies. An electromagnetic signal wave travels along a spring-shaped coil of wire, or helix, while electrons in a high-voltage beam travel through the helix at close to the speed of the signal wave. The electrons transfer power to the wave, which grows rapidly in power as it travels down the helix.

Transistors

There seems little doubt that transistors and other solid-state devices will gradually displace vacuum tubes even at higher frequencies and higher powers. Vacuum tubes made complicated signal processing possible. Transistors have made complicated signal processing cheap and reliable. At low powers transistors are much better than vacuum tubes ever were. Transistors amplify over broader bandwidths, use less power, cost less, are smaller, and have a seemingly endless life. Imagine a computer containing tens or hundreds of thousands of vacuum tubes. What a costly headache! Yet a pocket calculator contains tens of thousands of transistors and other solid-state elements. It lasts indefinitely. A simple flashlight battery or a solar cell can supply the power. You can buy a simple calculator for less than $10, and a complicated scientific calculator for a few tens of dollars. Most of the price pays for the case, keyboard, and perhaps the liquid crystal numerical display.

Think of the advances in the 100 years since Hertz's spark transmitter and the loop of wire with which he first detected electromagnetic waves. The intervening period has seen the rise and fall of the vacuum tube, a miraculous advance that lasted about 50 years. The past 40 years have witnessed the solid-state revolution. Progress is still so rapid that any particulars I might give would soon be out of date. The best I can do is to differentiate clearly among the physical phenomena involved.

In Hertz's spark transmitter the electric current crucial to the production of electromagnetic waves flowed erratically as a spark through the air gap. In de Forest's vacuum tube the electrons used for amplifying and other complicated functions had to be boiled out of a hot electrode into a vacuum before they could flow freely and be controlled. The vacuum was awkward, and the energy that was wasted in heating the cathode was usually far greater than the energy of the signal that was amplified or controlled. Shortly after I, John Pierce, came to Bell Laboratories in 1936 an older fellow worker, Myron Glass, told me that "nature abhors a vacuum tube." Oh, how true I found that to be.

In solid-state devices, electrons do their work inside semiconducting materials, where they occur naturally. No hot filaments; no vacuums. From this inspiration has come the seemingly endless solid-state revolution that began in 1947 with the invention of the transistor by John Bardeen, Walter H. Brattain, and William Shockley of Bell Laboratories. Three factors made the invention possible. One was the accumulating understanding of various aspects of matter that quantum theory had provided. Another was the preparation of very pure silicon for use as rectifiers and detectors or modulators in radar receivers during World War II. The third was Shockley's obsessive conviction that he could make a solid-state amplifier.

The invention of the transistor did far more than usher in a new age in communication. It inspired the basic research in solid-state physics that led to other practical advances, including light-emitting diodes used in optical fiber communication over short distances and the semiconductor lasers used in long-haul optical fiber communication. The invention of the transistor is one of those common instances in science in which a piece of applied research has led to a great deal of basic research. So had the invention of the steam engine and of the airplane in earlier days.

The invention of this marvelous device gave one of us (John Pierce) the opportunity to add a word to the English language. When Walter Brattain asked me what to call the new device, I told

CHAPTER 7

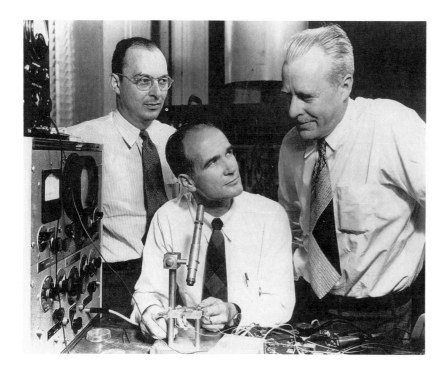

Left: John Bardeen, William Shockley, and Walter H. Brattain, pictured here in 1948 with apparatus they used in their experiments, won the Nobel prize for inventing the transistor. Right: The very first transistor.

him, "Transistor," It seemed logical enough. There were already Bell System devices called thermistors, whose resistance changed with temperature, and varistors, whose resistance changed with current. I was used to the ring of those names. Also, at the time we thought of the early point-contact transistor (then nameless) as the "dual" of the vacuum tube; in the operation of the two devices the roles of current and voltage were interchanged. The reasoning was simple. Vacuum tubes have transconductance; resistance is the dual of conductance, and transresistance would be the dual of transconductance—hence the name *transistor*.

Solid-State Quantum Physics

To understand how the transistor works we must take a look at the laws of quantum mechanics. We commonly picture an atom as a positive nucleus surrounded by orbiting electrons. Except in chemically inactive atoms, the outer, or *valence*, electrons are eas-

In 1972, these highly reliable British transistors for submarine cables won the Queen's Award for technological innovation. Although most transistors today are parts of integrated circuit chips, discrete, separate transistors are still used, usually as power output devices. The threaded portion of the transistor on the right screws into a metal structure in order to dissipate heat.

ily moved to a larger orbit, giving an *excited state,* or the outer electrons can be removed or new outer electrons added, giving an *ionized state.* Electrons can be in states of different energies—with different velocities and momenta. But only certain energy states are possible, each corresponding to a particular orbit of a valence electron. We can think of electron energy as measured in volts.

In some solids, atoms are arranged in an orderly, geometric fashion, called a crystal lattice. Atoms in such a solid share their valence electrons. Each allowed orbit or energy state is spread out into a band of energies. Rather than merely orbiting the nucleus, an electron can travel through the crystal lattice if its energy lies within such an energy band. Electrons can't travel with other energies; they can't *have* other energies. Corresponding to the different orbits that valence electrons can have, there are different bands of energy that allow free travel of electrons in a crystalline solid.

Vacuum tubes rely on the ability of electrons to travel freely with any energy through a vacuum. Transistors rely on the free travel of electrons through crystalline solids called semiconductors. Electron travel through semiconductors isn't quite as free as travel through a vacuum. As we have noted, electrons can travel freely only if their energies lie within certain ranges. Further, imperfections in the crystal lattice deflect electrons, and impurities can trap them. But, it is the almost free travel of individual electrons whose energies lie within certain bands or ranges of energy that is essential.

Each energy band consists of a large number of individual energies, or energy states. We can think of these states as coming in pairs of the same energy; one of the pair represents travel in a certain direction, and the other represents travel in the opposite direction. An electron in one state of a pair of states constitutes a current traveling in one direction; an electron traveling with the same energy in the other state constitutes an electric current traveling in the opposite direction. Thus, if there are electrons in *both* states, their motions produce no net electric current in either direction.

Semiconductors differ from true conductors, such as metals, in how full of electrons are the energy bands that allow free travel. In conductors, such as metals, one energy band that allows free travel of electrons is called the conduction band. In a strip or wire of such a conducting material, some electrons are freely traveling in one direction, and an equal number of electrons are freely traveling in the opposite direction; thus, there is no net flow of electrons, no electric current. The conduction band is only partly filled with electrons.

If we attach the terminals of a battery to the ends of a strip or wire of such conducting material, the battery can supply energy and move some electrons to higher energy levels. Some of the electrons that were traveling in one direction are turned around and put into higher energy levels, where they travel in the opposite direction. More electrons travel in one direction than in the other, and a current flows through the material.

In the energy bands of a pure semiconductor, the conduction band has no electrons in it. The next lowest energy band, the valence band, is completely filled with electrons, and as many are moving in one direction as in the opposite direction. There are no empty energy states in this valence band into which an electron might move to create a net current in one direction. Electrons cannot have energies above or below the conduction or valence bands; these ranges of energy are forbidden. Thus in semiconductors such as crystalline silicon, the conduction band is empty and the lower-energy valence band is completely full, so that while there are electrons in motion in various directions, there can be no net current. However, this situation can be altered by doping, the process of adding minute quantities of impurities.

Phosphorus has one more outer, or valence, electron than silicon. A minute trace (around a part per million) of such a so-called *n*-type material can contribute electrons to the lowest empty band. The formerly empty band turns into a conduction band, but unlike a metal, it contains only a few electrons. The silicon becomes a conducting *n*-type (for *n*egative electron) semiconductor. In the absence of an impressed electric field, half of the electrons in the newly formed conduction band will be going to the right and half to the left. But an electric field can cause electrons traveling to the right to go into slightly higher energy states and travel to the left, so an electric field can produce a current. The completely filled valence band contributes nothing to current flow.

An impurity such as boron, which has one less valence electron than silicon, cannot contribute an electron to the highest empty band. But it can abstract an electron from the valence band. For each atom in a *p*-type impurity, there will be one vacancy or hole in the otherwise full band. There can now be a net current flow in the band, but the current acts as if it were a moving positive charge—a moving positive *hole* where otherwise an electron might have been. A current appearing to consist of moving positive holes is the sign of a *p*-type semiconductor.

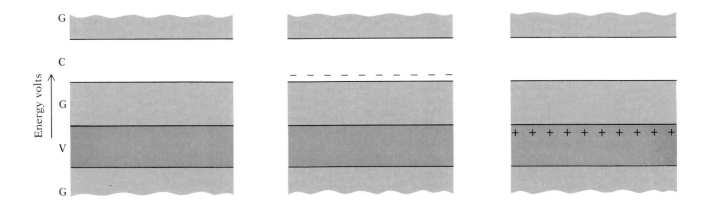

Energy volts

G

C

G

V

G

Left: In pure silicon, the conduction band C is empty, and the valence band V is filled with electrons. No net current flow is possible. Middle: An n-type impurity will add negative electrons to the conduction band, making possible a net current flow. Right: A p-type impurity will remove electrons from the valence band, leaving positive holes. The valence band can now conduct electricity. Electrons of different energies are shown at different heights. The diagrams are symbolic and do not represent physical structure.

Doping can produce different adjacent *n*-type and *p*-type regions in the same single crystal of silicon. In the *n*-type region, of electricity is conducted entirely by (negative) electrons. In the *p*-type region, electricity is conducted entirely by (positive) holes. Such a discontinuity between an *n*-type region and a *p*-type region is called a *pn* junction (for some reason, the *p* always comes first).

Electrons can flow across a *pn* junction from the *n*-type region into the *p*-type region, and holes can flow across a *pn* junction from the *p*-type region into the *n*-type region. This property makes transistors and integrated circuits possible.

A *pn* junction acts as a rectifier; it allows current to pass in one direction, but not the other. Suppose we make the *n*-type region positive with respect to the *p*-type region by connecting the negative pole of a battery to the *n*-type region and the positive pole to the *p*-type region. Then positive holes in the *p*-type region will be pushed away from the junction, and so will negative electrons in the *n*-type region. No current will flow. But, if we make the *n*-type region negative with respect to the *p*-type region by reversing the connections to the battery, electrons will be attracted into the positive *p*-type region, and holes in the *p*-type region will similarly be attracted into the *n*-type region. Negative electrons will flow from the *n*-type region into the *p*-type region, and positive holes will flow in the opposite direction from the *p*-type region into the *n*-type region. A net current will flow from positive to negative.

What happens to electrons that flow into a *p*-type region? They soon combine with the holes in that region. And holes that flow into an *n*-type region combine with electrons in the *n*-type region. In the combining of an electron and a hole, energy is emitted in the

CHAPTER 7

form of light. This is the basis of the semiconduction laser used in optical communication.

For reasons I won't discuss here, *pn* junctions are made asymmetrical; they are doped heavily on one side of the junction and lightly on the other side. Thus, there are almost always relatively few electrons on one side, or relatively few holes on the other.

MOS and CMOS

We are now in a position to understand the operation of the transistor. But, alas, there are many types of transistors. Here we will consider only the MOS (*metal oxide silicon*) transistor that is used in pocket calculators and home computers. There are two types of MOS transistors: in NMOS, the current carriers are (*negative*) electrons, and in PMOS, the current carriers are (*positive*) holes. In integrated circuit chips, the two MOS transistors are ingeniously combined in a CMOS (*complimentary MOS*) configuration so that during a computation a current flows only when a 1 changes to a 0 or a 0 to a 1. If no change, then no current, no power drain, and no heating of the transistors.

Transistors are built or deposited on a silicon surface called the substrate. In an NMOS transistor, the substrate is very weakly *p*-type. Two tiny, strongly *n*-type regions are produced on separate areas of the surface by doping. In operation, the right-hand *n*-type region is made positive with respect to the left-hand region, encouraging a current flow from left to right.

On the surface between the two tiny *n*-type regions is an insulating coating of silicon oxide. On the coating is deposited a metal *gate* electrode. Hence the description MOS (metal oxide semicon-

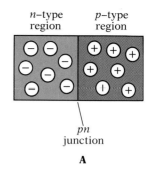

A

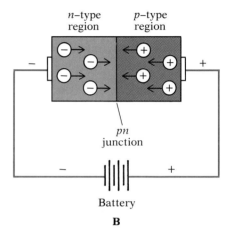

B

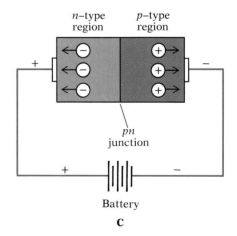

C

By doping adjacent tiny regions on the surface of very pure silicon with p-*type and* n-*type dopants, we can produce a* pn *junction (A). If the negative pole of a battery is connected to the* n-*type region and the positive pole of a battery to the* p-*type region, negative electrons will be attracted from the* n-*type region into the* p-*type region, and positive holes from the* p-*type region into the* n-*type region, and a current will flow (B). If, however, the positive pole of the battery is connected to the* n-*type region and the negative pole to the* p-*type region, both electrons and holes will be pushed away from the* pn *junction, and no current will flow (C). Thus, a* pn *junction allows current to pass in one direction, but not in the opposite direction.*

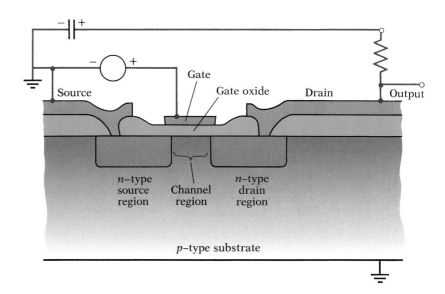

An NMOS transistor shown in cross section. The transistor is produced on the silicon surface of an integrated-circuit chip. Think of the size as measured in tens of micrometers. When no input voltage is applied between the electrode gate and the n-type source region, no current can flow to the n-type drain region, the output of the device. If, however, the gate is made positive, holes are pushed out of the p-type region between source and drain, and electrons are attracted into this region from the n-type source region. In effect, an n-type channel is produced between the source and the positive drain, and electrons can flow from source to drain.

ductor). The electrode brings in a signal from elsewhere in the computer.

When the gate electrode is at the same voltage as the left-hand or *source* n-type region, no current can flow to the positive right-hand *drain* region. The *pn* junction between the drain and the channel in a nonconducting state. Hence when no current enters the transistor from the electrode, no current leaves.

If an incoming current makes the gate sufficiently positive, holes will be driven from the channel region between source and the drain, and electrons will be drawn into this region from the source. In effect, the channel between the source and the drain will become *n*-type and can conduct current to the drain electrode. The NMOS transistor acts as an amplifier and control device. A small input current can produce a larger output current, sufficient to drive one or several other transistors.

Very Large Scale Intergrated Circuits

Today, transistors are largely parts of integrated circuits. Up to a few million transistors and other circuit elements are combined in an area usually a fraction of an inch square, making up an integrated circuit chip. Integrated circuits are produced on a surface of very pure silicon or, sometimes, gallium arsenide or other semiconductor by photographic, vapor-deposition, etching, and other

CHAPTER 7

techniques. Fifty or more chips are produced on one single-crystal, very pure disk of silicon (or other semiconductor material), 4 to 6 inches in diameter and about 15 thousandths of an inch thick. The integrated circuits on these chips are the heart of modern communication electronics.

The sound waves and fluctuating electric voltages that carry signals are continuously varying. We refer to such smoothly shifting quantities as *analog* to distinguish them from the abrupt on and off pulses of digital communication. Integrated circuits can be either analog or digital, but digital circuits predominate in number and complexity. Only digital circuits can be packed onto a chip capable of holding millions of transistors and other components.

We noted in Chapter 4 that pulse code modulation uses sequences of on and off pulses to represent analog signals as accurately as we may wish. When signals are in digital form, digital

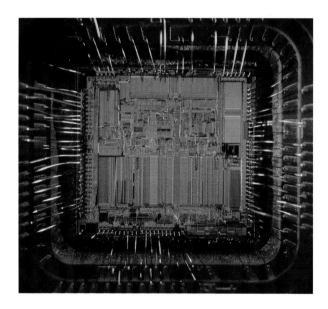

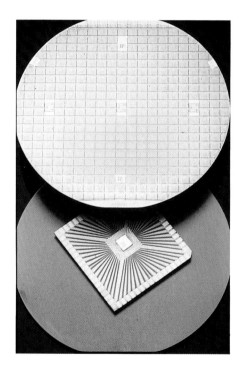

Left: A close-up of an LSI chip. Right: Integrated circuit chips are manufactured in quantity on single wafers 4 to 6 inches in diameter (top). Individual chips are diced apart from the wafer then mounted in dual in-line-packages (DIP) or in ceramic-based packages with connection pins for use in actual products (bottom). The transistors and other components are produced on the surface of wafer by photographic, vapor-deposition, etching, and other techniques.

integrated circuits can be used to transform, process, interleave, separate, switch, transmit, and receive them. In the digital domain, all the functions of communication are interchangeable, interconnectable—they are parts of one powerful art. Digital signals and large-scale integrated digital circuits are the wave of both the present and the future. We live, however, in an analog world. Our speech and motions are not on and off; they change smoothly and over a wide range.

The Digital and Analog Worlds

Sometimes a smooth, continuous motion can provide a direct analog-to-digital conversion. When we flip a switch or press a key, for example, the motion results in a desired digital effect. A light goes on or off, or a particular character appears in a particular place on a piece of paper or on a computer screen. More often, we wish to hear every inflection of a voice, or to see every tint of a picture. In such cases we must produce analog electric signals, which we can then convert to digital signals by means of special devices. At the receiving end, analog devices are needed to pick up and amplify weak signals that have become mixed with noise during transmission.

The solid-state art provides us with the devices necessary for generating, processing, and detecting analog signals. A number of these devices can make up circuits, which are themselves combined to produce analog integrated circuits. But, unlike the devices in LSI digital circuits, which need to respond only to on or off, the devices in analog integrated circuits must respond to and accurately pass on the wide range of voltages corresponding to an audio or video signal.

A communication system may use all digital circuits, or all analog circuits, or a combination of the two. Because of large-scale integration, digital circuits and digital processing are far cheaper than analog circuits and analog processing. Yet, even analog circuits and analog processing are far, far cheaper than the complicated devices that transmit and receive communication signals. Telephone transmitters (microphones) and telephone receivers are an exception. They are rather cheap; that is why telephony succeeded. In contrast, a printer, a display tube, a television camera, or a color picture tube are costly, and usually they consume a good deal of accurately regulated direct-current power, which is expensive to produce.

In the world of machines, the transducers and activators that feed signals to and from computers are far more costly and fallible than the computers that process the signals. Consider a modern car, in which the operation of the engine is monitored and continually adjusted for optimum performance. It is the transducers and activators required that are costly and fallible, not the computer. The same is true in flight control, traffic control, industrial processing—in practically everything.

The future of communication is clear. Everything will be put into a cheap, reliable digital form as soon as possible. Thus we can use cheap, small LSI digital circuits, and minimize the use of costly analog circuitry. But putting everything in digital form still leaves us with the costly interface between our analog world and the world of digital electric signals. This interface is a highly resistant challenge to science and technology.

Masers and Lasers

Any discussion of the devices on which communication depends would be incomplete without praise for the maser and the laser, two inventions of scientific interest and technological importance. Both of these devices are based on the laws of quantum mechanics.

Under ordinary circumstances, when an electromagnetic wave travels through matter, the molecules, atoms, or electrons absorb energy from the wave, and the wave steadily decreases in amplitude.

Quantum mechanics tells us that molecules, atoms, or electrons in solids occupy various discrete energy states or energy levels. If they are in a higher energy level and a lower energy level is available, they can fall to the lower energy level and emit electromagnetic waves. The frequency of the waves emitted is proportional to the difference between the initial and final energy levels, which can be measured in volts.

The presence of an electromagnetic wave whose frequency corresponds to the difference in energy can stimulate a fall from a higher to a lower energy level. The energy released by the fall is added to the energy of the electromagnetic wave, which grows in amplitude. This *stimulated emission* provides a mechanism for the amplification, or the generation, of coherent, single-frequency electromagnetic waves.

The trick is to get molecules, atoms, or electrons from a lower to a higher energy level. This can be done in a number of ways. By

A Semiconducting Laser Diode

In a semiconducting laser, electrons from the conduction band in an *n*-type region combine with holes from in the valence band of a *p*-type region. Energy produced through this process is added to an electromagnetic wave in the semiconductor and sustains the oscillation that produces monochromatic light. Silicon isn't a suitable semiconductor for this purpose. A suitable semiconductor must have a bandgap between the top of the valence band and the bottom of the conduction band that will produce electromagnetic waves of the desired frequency, or wavelength. The frequency of the radiation produced is proportional to the bandgap measured in volts. A few suitable semiconducting materials are listed in the table.

A semiconductor laser diode is a tiny strip a few micrometers on a side. As you look at the diagram, imagine the light to be coming out of the paper. In the active region, injected holes combine with injected electrons and transfer energy to the electromagnetic wave. The

Light-Emitting Semiconductors

MATERIAL	WAVELENGTH RANGE (MICROMETERS)	BANDGAP ENERGY (ELECTRON VOLTS)
Gallium arsenide	0.9	1.4
Aluminum gallium arsenide	0.8–0.9	1.4–1.55
Indium gallium arsenide	1.0–1.3	0.95–1.24
Indium gallium arsenide phosphide	0.9–1.7	0.73–1.35

using radiation we can drive the molecules in a solid to an energy level still higher, whence they fall to *the* higher energy level discussed above. Or we can apply a voltage across a *pn* junction in a semiconductor. Electrons will flow from the conduction band in the *n*-type region into the otherwise empty conduction band in the *p*-type region; there they can combine with positive holes in the lower valence band, emitting energy that causes an electromagnetic wave to grow in amplitude.

The first device to make use of this phenomenon was the maser, an acronym for *m*icrowave *a*mplification by *s*timulated *e*mission of *r*adiation. The maser provided an entirely new way to generate or amplify microwaves. The maser is associated with the active region is *n*-type aluminum gallium arsenide, with a bandgap of 1.55 electron volts. This bandgap will produce radiation with a wavelength of 0.9 micrometers. Above and below it are aluminum-gallium-arsenide layers somewhat differently doped to confine the wave to the central active region. Above and below these two confinement layers are gallium-arsenide layers that act as contacts so that an applied voltage can cause current to flow through the device. The layers above the active layer are doped *n*-type; they are connected to the negative voltage source and inject electrons into the active region. The layers below the active layer are doped *p*-type; they are connected to the positive voltage source, and inject holes into the active region.

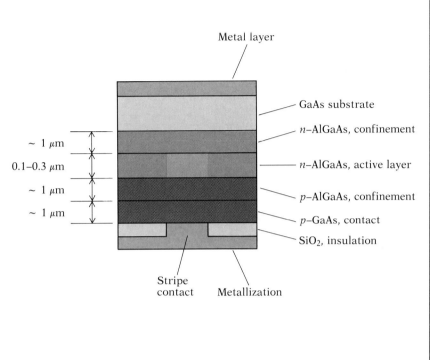

Metal layer

GaAs substrate

n–AlGaAs, confinement

n–AlGaAs, active layer

p–AlGaAs, confinement

p–GaAs, contact

SiO$_2$, insulation

~ 1 μm

0.1–0.3 μm

~ 1 μm

~ 1 μm

Stripe contact   Metallization

names of Nobel laureates Charles H. Townes of the United States and Aleksandr Prochorov and Nikolai Basov of the Soviet Union.

Masers add little noise when amplifying a weak signal, and therefore they are used as amplifiers for radio astronomy and for receiving signals from distant spacecraft such as *Voyager*. A sort of maser called the hydrogen maser is the most accurate clock in the world. The radio-frequency signal it produces is so steady that it is accurate to a few parts in 10^{15}—that is a millionth of a second in 30 years.

A laser (*l*ight *a*mplification by *s*timulated *e*mission of *r*adiation) is like a maser, but it operates in the frequency range of light. The American physicist Theodore H. Maiman designed the first

Dr. Theodore H. Maiman, inventor of the laser, with a synthetic ruby crystal —the heart of the laser.

laser in 1960. Lasers are our only source of coherent light waves. Coherent light waves are all of the same wavelength, and they are all in phase. The light waves produced by incandescent lamps, fluorescent tubes, and light-emitting diodes are incoherent. The light from such sources is a chaotic mixture of wavelengths.

If we transmit the incoherent, diffuse light from a luminescent diode through an optical fiber, it must go through a comparatively large fiber in many electromagnetic modes. As we noted in Chapter 6, different modes have different velocities and they reach a receiver at different times. That limits the shortness of a pulse, and hence the pulse rate that we can use.

The coherent, single-frequency light from a semiconductor laser can be focused into a fiber so narrow that it will transmit only one electromagnetic mode. Thus, by using coherent light and a very narrow fiber we can, using PCM, send very short pulses at a very high rate.

The Laws of God

In the very first chapter I quoted A. E. Housman:

> *The laws of God, the laws of man*
> *Let him keep who will and can*

We have pretty much finished with our exposition of the laws of nature, the possibilities that they afford, and the limitations that they impose. Along the way we have seen some of the effects of the laws of man.

There is a sort of middle ground between the laws of God— what nature allows—and the legislation and regulations imposed by governments—what man allows. This middle ground is the territory we call technology. The task of engineers is to produce things useful to man, or, at least, saleable. In such an endeavor, they are constrained by the laws of God—they can't achieve the impossible. They are also constrained by the laws of man, confining their achievements within the law. Acknowledging these limitations, engineers exercise varying degrees of ingenuity, and make various choices in accomplishing their ends.

Telephony affords a wonderful example of this middle ground. In advancing telephony, our nature and the nature of society may urge us in a general direction. But, we have a good deal of choice in what we set out to do and how we may accomplish it, however

CHAPTER 7

constrained we may be by nature's laws in carrying out our intentions.

As we go on to discuss how communication is organized and supplied, we should remember that the laws of nature *can't* be broken—they are the way the world works. It is through the laws of nature that *we* must work. Whatever our concept of communication may be—telegraphy, facsimile, telephony, television—how it *will* work depends on our knowledge of the laws of nature and the control that knowledge gives us over natural phenomena.

Switching and Signaling: The "Brains" of Communication

The thing that makes telephony so universal and unique is that it gives any one of us the ability to reach and communicate with any other. Its selective reach makes telephony very different from radio and television broadcasts, which are sent in all directions. If you want to pick up a radio or television signal, you have merely to buy a receiver and put up an antenna—and in some nations, pay a tax. The receiver decodes from the airwaves what is already going by, and you simply watch or listen to whoever is broadcasting. Cable television is not much different. Television signals are picked off the air, shifted in frequency in order to go through a coaxial cable and its amplifiers, and sent down the trunk of a branching, treelike network. Each branch of the network carries the same signal that is fed in at the trunk. The path is always between the broadcaster and the owner

The Gerrard exchange in London in 1926 required the labor of many operators, a common feature of all early telephone switching systems prior to electromechanical automation.

Number of subscribers	Number of on–off switches
2	1
3	3
4	6
S	$\frac{1}{2} \times S \times (S-1)$

The number of on-off switches needed to interconnect a large number of telephone subscribers quickly grows to an astronomical number.

of the set, and it is strictly one way. The same signal is sent to everyone and there is no way to send a signal back to the broadcaster, though many of us probably have on occasion shouted at our television receiver, and luckily have not been heard at the other end!

Telephony is deeply and radically different from radio and television, even though they all transmit and re-create audible sounds. Each subscriber has an individual circuit called a local loop, which goes from the subscriber's telephone to a centrally located telephone switching office. But the overall path traveled by the voice signals between one subscriber and another is different from call to call. Your subscriber loop can equally be a part of a circuit connecting you to your next-door neighbor or to a friend in Paris. Radio and television broadcast, and cable television as well, are mass communication from the few to the many. Telephony, like mail service, provides individual, person-to-person communication. To achieve that person-to-person service, transmission paths must be rapidly and drastically reconfigured, and it is switching that makes that flexibility possible.

Manual Plug-and-Jack Switching

The problem of arranging for any telephone subscriber to talk to any other subscriber has plagued telephone switching from its first days. One approach would be to provide a separate path between all possible pairs of the subscriber loops that come to the switching office. Each path would have its own on-off switch; turning on a switch would connect the two subscribers on that path. Although possible in theory, there is something fundamentally wrong with that approach. Imagine a very small office serving only four subscribers. Each subscriber would require a separate switched path to each of the other three subscribers. Since there are six links, or paths, connecting the four subscribers, six on-off switches would be needed to connect one subscriber with any of the other three. The simple equation for the number of links or paths, W, connecting S subscribers is $W = \frac{1}{2} \times S \times (S - 1)$. Thus, a switching office serving 10,000 subscribers, a modest number in countries with tens of millions of telephones, would require $\frac{1}{2} \times 10,000 \times 9999$, or nearly 50 million, on-off switches.

There must be a better way, and, indeed, there is. I think we can best gain insight into how switching is done by considering the

By plugging in a cord, the operator could converse with the calling party. The other end of the cord then connected the calling party with the called party.

manual plug-and-jack switchboard. It is now all but extinct in many nations, although it still exists in some countries that are less advanced technologically. The manual switchboard with its human operator, however, could offer a high level of individualized personal service that was lost with automated electromechanical switching; more about that story later.

Each switchboard operator sat in front of a shallow desk containing a row of about 18 pairs of plugs used to make connections.

Each pair of plugs consisted of a front plug and a back plug connected together by electrical cords. In front of each pair of plugs was a listening key and a ringing key. The plugs could be inserted into jacks on the vertical panel behind the desk. At the bottom of the panel were about 120 answering jacks with little signaling lamps above them; when a subscriber took the receiver off the hook, one of the lamps lit up. The operator put the back plug of a pair of plugs into a jack, pushed the listening key, and said, "Number please." On the upper part of the board, subscriber multiple jacks (as many as 10,000 on some boards) were within each operator's reach; jacks to the side could be shared by adjacent operators. After hearing the number, the operator touched the tip of the front plug to the sleeve of the appropriate jack. A click meant that the line was busy; if no click was heard, the operator inserted the plug and pushed the ringing key. When either subscriber hung up, a lamp associated with the pair of plugs lit up, and the operator pulled the plugs out. The lamp associated with the forward plug also enabled the operator to determine when the called subscriber answered the phone.

Using a plug-and-jack switchboard, a single human operator with 18 electrical cords can answer any of 120 subscribers and can connect any of them to any one of as many as 10,000 subscribers. If advanced switching technology had not been developed and the telephone company still had one operator for every 120 of some 100 million telephones, it would take at least 800,000 operators (or more than 2,400,000 on three shifts)—a sizable work force. In fact, the whole telephone system in the United States employed only about 800,000 people in all a few years ago, and they each did not work 24 hours a day.

The plug-and-jack switchboard provided a means for handling 10,000 subscribers with far less than 50 million switches. But there were more than 10,000 subscribers even in the days of plug-and-jack switchboards. What if the call were for another switching office? On the switchboard, above the set of answering jacks and below the set of subscriber multiple jacks, were a number of trunk jacks. They led to trunks, which are circuits that connect wire centers to one another. The trunk jacks enabled the operator to reach an operator at another telephone office and pass the telephone number on so that the call could be completed.

In a telephone office with manual switching, each operator could reach and connect a call to all 10,000 lines but could not answer all incoming calls, and each operator's 18 pairs of jacks represented only a fraction of the traffic through the office. The pattern of 10,000 jacks serving the same 10,000 subscribers was repeated over and over at different operator positions.

Manual switching illustrates the functions that any switching system must perform: ascertain when a subscriber wants service ("goes off hook"); receive the telephone number; test to see if the called number is busy and inform the caller if it is busy; make a connection; ring the called party; and go back to the initial (disconnected) state when one party hangs up.

Manual switching also illustrates multiprocessing, which has been important in automatic switching. One operator cannot handle all calls. Therefore, several operators are used to perform essentially identical tasks, and they share the total work of switching.

Manual switching illustrates blocking, as well. There are only so many pairs of front and back plugs in an office. If all are in use, no more calls can be made. Another kind of blocking can prevent a call from reaching another switching office. There are only so many trunks from one office to another. If they are all in use, no more calls can be made between the two offices. To this day you sometimes get a busy signal, not because the phone you have tried to call is in use, but because there are insufficient circuits to handle all calls, and your call has been blocked.

We have noted that a single operator was able to reach only 10,000 subscribers. But, an operator could reach far more subscribers by forwarding a call through a trunk line to another operator at a different switching office; the call thus went through two stages of switching. As telephone service grew in large cities, it was clearly inefficient to provide a completely separate switching office for each 10,000 subscribers; too many buildings, each serving only 10,000 lines, would have to be scattered about the city. Instead, two stages of switching were built into large, tandem manual switching systems that could serve more than 10,000 lines from a single building. The operator at an A position answered the subscriber, took the number, and plugged into a trunk going to an operator at a B position, who could reach a jack corresponding to the number called. The operator at the A position passed the number on to the operator at the B position, who completed the call.

The Solution to Blocking

As multistage switching, within tandem offices or between offices, connected more and more subscribers, engineers began to study blocking and efficient design. Suppose we vary the number of stages of switching used in handling a given number of telephones with a given amount of traffic. How much equipment and how many operators are needed for each number of stages? How can a given amount of equipment and personnel serve the largest number of subscribers?

In addressing the general problem of blocking, the simplest question we can ask is: If we have M possible callers and only N circuits (for example, pairs of plugs), what is the probability that a caller will be blocked? If we knew exactly when each caller would attempt a call and exactly how long the completed call would last, we could laboriously work out the exact fraction of calls that are blocked. Because we do not have such information, that approach is useless. We must make assumptions concerning how many times a day a subscriber will try to make a call and how long the call will last, and so a solution to the blocking problem will give the probability that a specified number of calls will be blocked. We need a simple probabilistic mathematical model of calling behavior.

Important early work concerning blocking was done by M. C. Rorty of AT&T in 1903 and by E. C. Molina, also of AT&T, in 1908. Molina was the first to assume that the probability that a given number of calls will be initiated in a given time period has what is called a Poisson distribution. From telephone records, engineers know what the average number of calls in a given period of time is. But the actual number of calls during that time period will often differ from the average. The Poisson distribution gives the probability that a specified number n of calls will be dialed in that period, for different numbers n, and so gives the probability of having x calls concurrent in that period. According to the Poisson distribution, if the average number of calls that are dialed in one second is a, the probability that n calls will be dialed in a particular second, which we call $P(n)$, is $P(n) = a^n e^{-a}/n!$. The distribution follows from the very simple assumption that a caller is equally likely to initiate a call at any time. In more mathematical terms, the probability that a call will originate in an exceedingly short time interval is simply a constant times the length of the short time interval. If the short time interval really is zero, no calls will originate in it. Zero is a limit. For any finite time interval there is a chance that one or more calls will originate in it, and the chance for any number n of calls to originate is greater than the length of the interval.

In 1918, the Danish telephone engineer A. K. Erlang published his work on blocking in *The Post Office Electrical Engineers' Jour-*

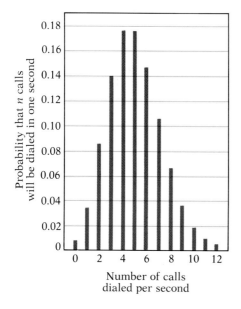

The probability that a given number of calls will be dialed in a given time period has a Poisson distribution. The histogram shows the probability that n *calls will be dialed in a particular second if the average number of calls that are dialed in one second is five and the number* n *of calls varies from 0 to 12.*

CHAPTER 8

nal, a British publication. Like Molina, Erlang assumed a Poisson distribution of calls arriving in a given time interval. Molina, however, had assumed a constant holding time, or duration, for all calls, whereas Erlang assumed an exponential distribution for holding times. An exponential distribution means that longer calls occur less frequently than shorter calls. Erlang's assumption is not only closer to the facts, it is overwhelmingly simpler. It means that the number of calls that terminate in any short period of time is proportional to the length of the period of time and to the number of calls in progress and does not depend on when these calls were initiated. In other words, for Erlang's exponential "decay," the chance of a call falling out of a given time interval is independent of when the call started. Furthermore, whereas Molina assumed that blocked calls are held for the same constant duration as completed calls and are only then lost if no server capacity becomes available, Erlang assumed that blocked calls are immediately cleared and lost and do not return. A formula that Erlang worked out based on these assumptions (Erlang B) is still in use in telephone engineering.

In honor of A. K. Erlang, the fundamental unit of telephone traffic is called the Erlang. The traffic measured in Erlangs is the number of call hours per hour. A single voice circuit fully occupied for one hour carries one Erlang of traffic for that hour. If the same voice circuit carried only a single 15-minute call in an hour, the traffic would be 0.25 Erlang.

Usually traffic is measured for a collection of voice circuits between two offices or for a number of subscriber lines served by a switching system. Essentially, the traffic in Erlangs is the average number of simultaneous calls served by some specific telecommunications facility in an hour. The chosen hour usually is the busiest hour but also might be an average over several hours.

In designing a telephone system, engineers must determine how many circuits, or how many inputs and outputs on a switch-

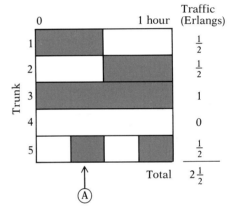

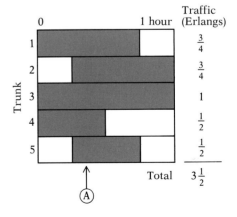

Five trunks connecting two local offices carry differing amounts of total traffic in one hour, $2\frac{1}{2}$ Erlangs for the five trunks on the left and $3\frac{1}{2}$ Erlangs for the trunks on the right. A call arriving at time A on the left would find idle capacity and would be served. The five trunks on the right carry more traffic and are more fully occupied. A call arriving at time A would find all trunks occupied and could not be served. Actual calls are much shorter than those used in these examples.

ing system (called ports), are needed to handle a known amount of traffic and to maintain a desired probability of blocking. The probabilistic models devised by mathematicians like Erlang and Molina give the solution under certain assumptions. They usually predict the traffic carried in Erlangs for different probabilities of blocking and different number of trunks (or ports) in the serving group.

If we compare the two switching networks illustrated on this page, we see how efficient design can reduce expense. The simplest

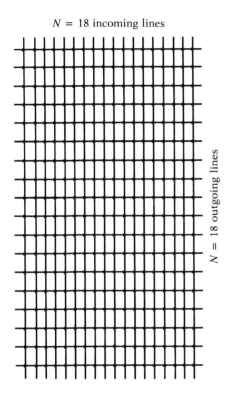

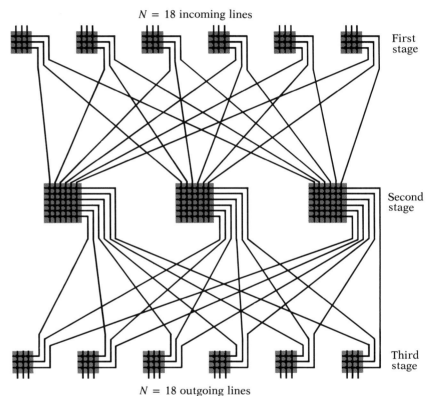

An 18-by-18 matrix (left) serves 18 incoming lines and 18 outgoing lines in a strictly nonblocking manner. It requires 18^2, or 324 switches, represented as the dots at the crosspoints in the matrix. A three-stage switching network (right) requires only 216 switches, but the paths through the switching network might need to be rearranged to prevent blocking.

switching network connects N incoming lines to N outgoing lines in an N-by-N matrix, where every line crosses every other line and a switch is placed at each intersection (left). The number of switches in such a network is N^2. In the example, N is 18, and the number of switches N^2 equals 18^2, or 324. Fewer switches are needed, however, in a multistage network. In the three-stage network (right), the first stage connects 18 incoming lines to six 3-by-3 matrixes, each matrix having three incoming lines and three outlets. In a similar way, the third stage connects 18 outgoing lines to six 3-by-3 matrixes, each having three inlets and three outgoing lines. The second stage consists of three 6-by-6 matrixes. The number of switches in this three-stage network is 54 for the first stage, 108 for the second stage, and 54 for the third stage, with a total of 216 switches for the whole switching network, clearly a saving over the 324 switches required for the 18-by-18 matrix.

We could construct a second three-stage network with even fewer switches. It would consist of three 6-by-2 first-stage matrixes, one 6-by-6 second stage matrix, and three 2-by-6 third-stage matrixes. The total number of switches would be 108.

All three networks we have discussed connect 18 incoming lines and 18 outgoing lines. The 18-by-18 matrix is nonblocking (any incoming line may be connected to any unused outgoing line regardless of traffic in between). The first three-stage network is also nonblocking, but the paths through the network might have to be rearranged to serve all the required connections. The second three-stage network is blocking; as an example, if three incoming lines to the first 6-by-2 matrix required service, only two could be served since there are only two outlets. Some degree of blocking can be acceptable if all incoming lines need not be served simultaneously. Switching engineers calculate the probability of blocking for different switching networks and seek a compromise between an acceptable probability of blocking and the number of switches.

Automatic Switching

Manual switching required a large number of human operators and was slow. Today switching is automatic. A telephone network of the present size would be impractical without automatic switching systems. But the basic ideas and some of the problems of manual switching persist. A switching system still performs two sorts of functions. The control part performs the functions of the operator. The switching network plays the role of the plugs and jacks in establishing the physical talking path. Nearly all switching networks are multistage networks, carefully designed to give acceptably low blocking without using an excessive number of switches.

Present-day automatic switching systems have much of the flexibility and clear organization of plug-and-jack manual switching. Today's automatic systems, however, are the result of a long road of development, and early automatic switching was very different. The control function was mixed up with the switches that provided the talking path.

No one thought that automatic switching would solve all problems. The early engineers were looking for some satisfactory means of speeding up the switching process and preventing an intolerably costly growth in the number of telephone operators. In 1891, Almon B. Strowger, an undertaker from Kansas City, Missouri, patented an automatic switching system that was actually constructed in 1892 and operated for four years. The first practical automatic switching, called step-by-step (or simply step) switching, grew from Strowger's ideas. In 1896, associates of Strowger invented the familiar rotating fingerwheel dial, which allowed callers to give instructions to the switching system without the intervention of a human operator. The Automatic Electric Company was founded by Strowger's associates to manufacture the automatic switching systems and dial telephones.

The basic step switch used in telephony has a movable pair of metallic contacts, or wipers, attached to a vertical rod, and 10 semicircular levels of fixed metallic contacts, with 10 pairs of contacts on each level. The movement of the switch is controlled by dial pulses. A succession of 1 to 10 dial pulses steps the vertical rod up to the designated level, and a second succession of 1 to 10 dial pulses rotates the rod so that the movable contacts engage the designated pair of fixed contacts. Thus when a movable contact is stepped up and rotated, it can engage any one of 100 pairs of fixed contacts. In a 100-line exchange using a step switch for each subscriber, one switch operated by two successions of dial pulses can connect any subscriber to 100 different lines. Unfortunately, the step switch cannot make a connection without a large amount of mechanical movement, and the sliding action across the contacts leads to much wear and tear and electrical noise.

In larger switching offices, the first step switch connects the subscriber to a second switch and so on until the subscriber finally reaches the desired line. For example, to connect one subscriber to

In a Strowger switch, the actual connection is made at the bottom of the switch, where movable wipers attached to a vertical rod touch one of the stationary contacts. The relays at the top of the switch interpret dial pulses and control the vertical and rotary movement of the vertical rod.

CHAPTER 8

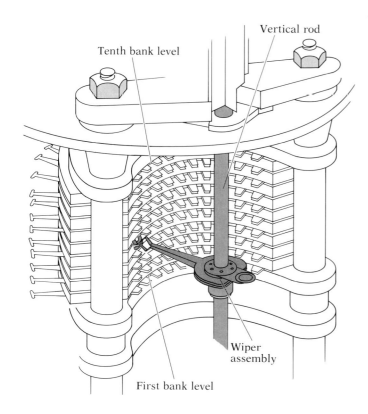

Tenth bank level

Vertical rod

First bank level

Wiper assembly

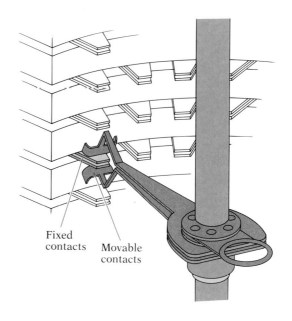

Fixed contacts

Movable contacts

The movable contacts in a step-by-step switch can connect to any of a 100 different pairs of fixed contacts, each leading to a different line.

one of 10,000 (numbered 0 to 9999) other subscribers through a step-by-step switching system requires the dialing of four digits that control the operation of four stages of step switching. When the calling subscriber goes off hook, the flow of direct current is sensed and causes the line finder to hunt vertically and then horizontally to connect itself to that subscriber's line. Dial tone is then provided to the subscriber. The line finder is wired to the first selector switch. The subscriber dials the first digit, which causes the first selector switch to step vertically the dialed number of pulses and then hunt horizontally for an idle line to the next selector. The second dialed digit operates the second selector switch, which likewise steps vertically and then hunts horizontally for an idle line to the connector switch. This next switch actually makes

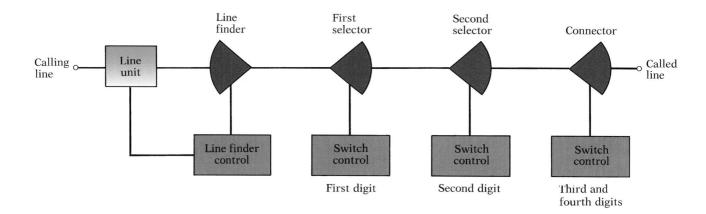

Line finder   First selector   Second selector   Connector

Calling line   Line unit   Called line

Line finder control   Switch control   Switch control   Switch control

First digit   Second digit   Third and fourth digits

After the line finder has connected the caller to a step-by-step switching network, the signal is sent through a series of switches until it reaches the called subscriber's line.

the final connection to the desired party; it is operated by the third and fourth dialed digits. The third dialed digit causes the connector switch to step vertically, and the fourth, and last, dialed digit causes the connector switch to rotate horizontally, thereby finding the contacts that lead to the desired party. If the called party's line is busy, a busy tone is returned to the calling subscriber; otherwise, the called party's telephone is rung.

The step switches are controlled by relays that are part of each switch. A relay is an on-off switch that opens or closes in response to an electric current flowing in another separate circuit. The control is thus used to make the connection and cannot be used for any other call. Step switching is inefficient because the control function (which establishes the connections) and the switching network (which maintains the connections) are intermingled.

Strowger's step-by-step automatic switching systems were sold to and used by the non-Bell, or independent, telephone companies in the United States and by telephone companies in other countries. The Bell companies waited until 1919 before adopting Strowger's invention, in part because they did not believe that it was desirable for subscribers to dial telephone numbers themselves, thereby directly operating the switching machines. Instead the Bell companies continued to use human operators to make the connections. This is one of the few examples where the Bell companies were slow in adopting new technology.

Common Control

Step switching proved to be cheap and reliable, but it has inherent limitations. One trouble is that its brains are all mixed up with its brawn. In manual switching, there is a clear division between the function of control that the operator exercises in setting up a connection and the switching network (plugs and jacks) that maintains the connection until it is again taken down by the operator. In step-by-step switching, control and the switching network are intermingled in the complicated step switches and the relays that control them. And the costly control equipment that sets up the talking path is used not simply while a call is being set up but for the whole length of the call. For many years, step switching was the cheapest automatic switching available. But step switching was ultimately replaced by common-control switching systems, which finally became more economical. Today it is nearly impossible to find step switching in the United States and in many other countries.

In common-control switching systems, the clear separation between control and the switching network is reestablished. The complicated control apparatus that replaces the telephone operator is used only in setting up a call (or terminating it, which is far simpler) and is then available to serve another subscriber. Because the control apparatus is used for only a short time during each call, we can afford to make it more complicated and flexible. And common control has certain inherent advantages.

Early common-control switching networks were electromechanical: they were made up of relays and other switches that were operated by magnets or motors. The general principle of common control survives in electronic switching systems. When the subscriber goes off hook in an electromechanical common-control switching office, the subscriber's line is connected to an originating register, which "listens to" and "remembers" the dialed number by storing all the digits of the number dialed. To connect the calling party with the dialed party, a sequence of switches in the switching network must be closed. A device called

Common control equipment

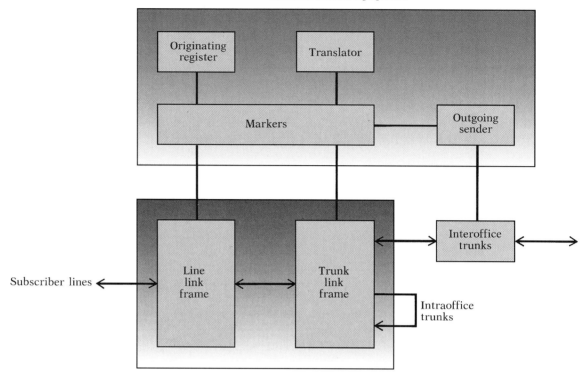

Switching network

The crossbar switching system consists of an electromechanical switching network under the control of common control equipment that is used during the set-up of a connection and then is released to be used for other calls. The "intelligence" in the common control equipment is in devices called markers. A calling party is connected by a dial-tone marker to an idle originating register that supplies dial tone and then receives and stores the dialed digits. A completing marker examines the dialed number to determine whether it is an intraoffice call, toll call, or call to another local office. If it is an intraoffice call, the dialed number is translated to a location on the line link frame and a marker sets up a path through the switching network to complete the call. If the call is outside the office, the called number is sent down the trunk by an outgoing sender. The switching network consists of a line link frame and a trunk link frame. Both the line link frame and the trunk link frame are two-stage switching networks using electromechanical crossbar switches, usually with 20 inlets and 10 outlets.

a marker chooses the switches to be closed. After examining the stored number, the marker finds an unused path through the switching network, from caller to called, and then produces signals that close the appropriate switches. If a marker determines that the dialed number is a call to an outside telephone exchange, an idle interoffice trunk is seized and a device called a sender sends the called number down the trunk.

Instead of using step-by-step switches, a common-control switching network contains interconnected matrix switches. A matrix switch consists of crosspoints between input and output lines. Simple on-off switches at the crosspoints make the connections. Many simultaneous paths through each switch are possible. Electromechanical systems often use crossbar switches having 20 inputs and 10 outputs. The Bell System's No.1 crossbar switching system was first installed in 1938. New crossbar switching systems were still being innovated as late as 1974 when the first installation of AT&T's No.3 crossbar system for rural areas was made. At

Electrochemical switching systems reached their zenith with crossbar switching. The bays and bays of identical crossbar switches of this No.5 crossbar system are typical of these systems.

On–off

Rotary

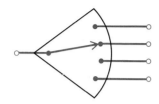

Matrix

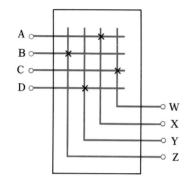

Signals can be switched in physical space by a variety of different ways. The simplest is an on-off switch, such as turns electric lights on and off. It is depicted as two dots and a line. The dots represent the contacts, and the line represents the means for completing the electric circuit. The on-off switch is usually shown in the open position with no current flowing. A rotary switch connects one contact to one of many. The line with the arrow at one end depicts the means for connecting the contacts.

The matrix switch consists of crosspoints between input and output lines. The example shows four input lines simultaneously connected to four different output lines. The small × depicts actual crosspoint connections, and in this example input A is connected to output X; B to Z; C to W; and D to Y.

that time, only about one-third of Bell System lines were being served by step switching systems. The remainder were being served by crossbar and by the ever-growing new electronic switching systems.

To find ways to exploit the flexibility of common-control automatic switching, engineers have done deep mathematical studies of multistage switching networks. In these studies, they experiment interconnecting the various stages of switching by a limited number of transfer trunks, or links. The problem of blocking is more complicated in such multistage switching systems than in the one to which the Erlang B formula applies. Indeed, a multistage switching network can be designed to be nonblocking, as demonstrated in 1953 by Charles Clos, a Bell Laboratories engineer. Truly nonblocking networks would contain too many switches, however, to be practical. It is better to use a less costly network in which the probability of blocking can be shown to be small. For example, A. C. Jacobaeus of the Ericsson company in Sweden published an important paper in 1950 that showed how to design such multistage networks.

Translation

Because the entire dialed number is registered, or held, in common-control switching, a sort of translation becomes possible that is of great value to both telephone users and telephone companies. In common-control switching each subscriber has a number that is determined by the physical location on the final switch of the terminals leading to the subscriber's line. This number is called the office number, because the office switching equipment uses this number to find the correct path through the network of switches. However, the telephone company assigns an entirely different telephone number for people to dial when calling the subscriber. Because that number is listed in the telephone directory, it is called the directory number. A piece of equipment in the switching system called the translator changes the directory number into the office number so that the party's line can be found. After the calling party has dialed the number into the sender, the translator performs the required translation from directory number to office number. A subscriber can retain the same directory number even if the subscriber moves and the subscriber loop is connected to a different pair of terminals in the switching office. Before any switches are operated, a common-control system can recognize area codes, no-charge 800 numbers, and other special telephone numbers.

Translation has advantages for the phone company as well as for the subscriber. Multistage switching has a lot of first-stage switches that can be connected by a limited number of paths to a lot of second-stage switches, and so on. Blocking is least if a mix of high-usage and low-usage subscribers is connected to each first stage. Translation allows such a distribution of subscribers among first stages, independent of the telephone numbers assigned to the subscribers.

Electronic Control Dawns

In early common-control switching, control was exercised by what we would now describe as a hard-wired, multiprocessor, special-purpose relay computer. *Hard wired* means that the computer is "programmed" by means of wires soldered to and interconnecting appropriate terminals of various apparatus. Because one mechanical processor cannot work fast enough to handle all calls, several identical devices are provided for each function, hence the term *multiprocessor*. The devices are line finders to seek out a customer who wants service; senders to register the digits of the called number and transmit the called office code to a marker; and markers to receive code digits from the sender, select an idle path through the switching system, and return information to the sender concerning details of the call. The common-control computer is termed *special*

Technicians determine the status and potential troubles of an AT&T No.1A ESS™ switching system from its maintenance control center.

purpose because it is designed to close appropriate switches when a person makes a telephone call, rather than to multiply numbers or sort data, like a general-purpose computer. Finally, *relay* refers to the devices that actually make up the computerlike mechanism that controls the switches.

When the transistor and the magnetic-core memory made general-purpose computers reliable and reasonably inexpensive, there was a strong push toward a concept that had not proved feasible in the days of the vacuum tube—electronic switching. The original idea was a switching system without metallic contacts, or with very few. In the first large-scale electronic switching system, No.1 ESS™, however, the talking path is made up of special metallic switches that can be opened or closed by a pulse of electric current. The operation of the switches is controlled by what is essentially a general-purpose electronic digital computer.

In No.1 ESS, a metallic talking path was chosen primarily because it fitted in with existing telephone practices and equipment, including the subscriber's telephone set. A telephone bell rings, for example, when a 20-Hz signal of about 90 volts is applied to the subscriber loop. That voltage would destroy most transistor circuits, but it will not damage metallic contacts such as those in telephone switches. The carbon transmitter in the subscriber's set, which acts as an amplifier and microphone, needs an electric current of about one-twentieth of an ampere to operate best. That current is high by transistor standards. Indeed, it represents a power far higher than is needed to produce the desired sound level in a telephone receiver. The voltage applied to a subscriber loop at the central office is 48 volts, which is also high by transistor standards.

In addition, the two-wire nature of local telephone transmission means that one must be able to talk both ways through the switching network without substantial loss of signal power. This is possible with metallic contacts. However, if the switching network itself is to be electronic, there must be a low-loss, symmetrical, two-way electronic switch. Alternatively, two-wire to four-wire conversion between the subscriber and the electronic network could be used, with amplifiers added to supply an adequate signal. No.1 ESS avoided the dilemma by using metallic contacts controlled by an electronic computer. The control functions were programmed into the computer rather than hard wired into a network of devices, which meant that various features could be added or changed without soldering or unsoldering wires. It also meant a formidable task of writing and rewriting large computer programs.

CHAPTER 8

When No.1 ESS was designed, multiprocessing was no longer essential. The computer could work fast enough to perform every function for all callers. To ensure reliability, the system used two complete control computers, both always turned on and checking each other. Putting all functions for all callers into the program of one general-purpose computer required a computer program unparalleled in both reliability and complexity. As computer logic and memory have become cheaper, special-purpose electronic hardware has been used to perform simple, repetitive functions, so that the central computer in the control system handles complicated functions and audits itself to see if it is working properly.

After some initial problems, No.1 ESS proved to be a great success. It occupies less space than electromechanical systems, uses less power, and requires less maintenance. Program changes have provided new and updated services. No.1 ESS has been followed by other switching systems that use electronic computers to control mechanical switching networks made up of metallic switches.

Time-Division Switching

The trouble with combining electronic control and mechanical switching is the cost of the electromechanical switching network itself. With the advent of large-scale integrated circuits and new forms of memory, the cost of the brainy computer has fallen rapidly, while the cost of an electromechanical network of switches has scarcely changed at all. It is clear that if the cost of switching is to be brought down drastically, both the control and switching functions must be electronic and must, indeed, ultimately be carried out by large-scale integrated (LSI) circuits. Large-scale integration is particularly well suited for handling on-off signals and this points toward the use in switching of pulse code modulation, in which the voice signals are transmitted as 8000 groupings of on-off pulses, or bits, per second. Each grouping consists of 8 bits, and thus the overall rate of transmission is 64,000 bits per second.

In PCM many messages are sent over one communication channel in different, repetitively occurring time slots. A new kind of switching is needed to take advantage of PCM. Called time-division switching, it stands in contrast to space-division switching, as used in such switching systems as step-by-step, crossbar, and No. 1 ESS.

Space-division switching provides distinct and electrically separable talking paths for the full duration of each call in progress. Thus in an eight-stage network, each call ties up eight talking paths. In the time-division alternative, many calls can go over the same path, but at different times. The path, called a bus, is accessible to many parties. Switches connect the parties of one call to the bus. The switches are closed just long enough to send eight pulses; then other switches are closed to connect other parties to the bus and send eight pulses; and so on. Each party gets the bus 8000 times per second.

Let us compare extreme forms of space-division and time-division switching. In a one-stage space-division system serving 10,000 subscribers, about 50 million on-off switches would be required to connect any party to any other party. In a simple time-division switching system serving 10,000 parties, some 10,000 switches would be sufficient. Switches would be closed in pairs to connect two parties to the bus at once.

People have known of the economy of time-division switching for a long time. The question was how to attain such economy in a practical system. An experimental time-division PCM switching system was built at Bell Laboratories as early as 1958. Indeed, the one substantial contribution to telephone switching made by one of the authors (John Pierce) was to persuade the people in his division to drop work on electronic space-division switching and concentrate all their efforts on time-division switching. Some early time-division systems actually handled real telephone calls. In the early 1960s, an electronic digital switching system built at the University of Tokyo under the direction of Hiroshi Inose handled part of the telephone traffic in the electrical engineering department.

Time-Slot Interchange

Time-division switching uses a digital, or PCM, representation of the voice signal. The analog voice signal is sampled at twice its bandwidth, or 8000 times per second. Each sample is then quantized into one of a fixed number of values, usually 256, called levels. Each level is encoded as a binary number of 8 bits. Finally, the bits are sent as a stream of on-off pulses; each grouping of eight pulses represents a sample value of the analog voice signal and is called a time slot.

Time-division multiplexing combines together a number of PCM voice channels. Sample values from each channel in turn follow in sequence along the transmission path. If, for example,

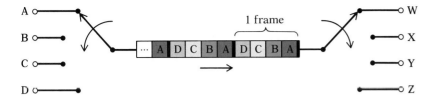

A single time-division multiplexed trunk serves the four voice circuits A, B, C, and D, which are connected to the four other voice circuits W, X, Y, and Z at the receiving end. A rotary switch samples the four input circuits in sequence, and the four sample values are transmitted in sequence along the trunk. At the receiving end, another rotary switch distributes the sample values in sequence to the appropriate terminal circuits.

128 PCM circuits are time multiplexed together, 128 groupings of 8 bits each, one grouping from each circuit, are sent down the line in $\frac{1}{8000}$ second. The 128 groupings, or time slots, are called a frame. At the receiver, each time slot with its grouping of 8 bits is transferred to a separate channel.

Blocking in time-division switching presents a somewhat different problem from blocking in space-division switching. For a call to get through a time-division system, free time slots must occur at each stage of switching, and the free time slots must occur at the same times. Inose's invention of time-slot interchange was an important advance in reducing blocking in time-division switching.

If time slots are interchanged along the transmission path, then input circuits will feed into different output circuits. The interchange of time slots is a form of time-division switching and is performed by a device called a time-slot interchange unit (TSIU), which consists of a temporary memory, called a buffer. Sample values enter the buffer in sequence, but they are shortly read out in a different order. In this manner, time slots are reordered, and the sample values in the reordered time slots are connected to different circuits. The buffer memory is solid-state, and transistorized switches enter and read the bits from the memory, all under the control of a processor.

It would be impractical to design a time-division PCM switching system that consisted of only a simple time-slot interchange unit. A practical time-division switching system operates in stages, with some stages using time-division switching and one or more

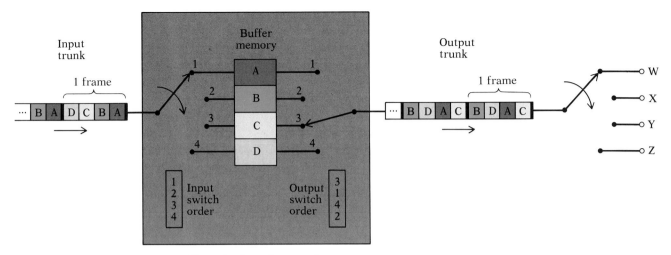

Input trunk

1 frame

··· B A D C B A

Buffer memory

1 A 1
2 B 2
3 C 3
4 D 4

Input switch order
1 2 3 4

Output switch order
3 1 4 2

Output trunk

1 frame

··· B D A C B D A C

W
X
Y
Z

Time–slot interchange unit

A time-slot interchange unit (TSIU) reorders the time sequence of the sample values sent along the trunk. The sample values are entered in order into a buffer memory in the TSIU by a rotary switch that steps through positions 1, 2, 3, and 4. The sample values are then read out of the buffer by a second rotary switch whose order of stepping can be changed to accomplish a reordering of the time slots. If the first time slot is designated for circuit W, the second for circuit X, and so forth, then time-slot interchange has connected circuit A to circuit X, B to Z, C to W, and D to Y.

other stages using space-division switching. The time-division switching is accomplished by time-slot interchange. The space switching is performed by a high-speed, electronic matrix switch that reconfigures itself rapidly to switch samples from one time-slot interchange unit to another. A space-division switch used this way is called a time-multiplexed space-division switch (TMS).

The diagram on the facing page illustrates a simple time-division switching system. Digital trunk A serves four digital lines, A1, A2, A3, and A4; digital trunk B serves four digital lines, B1, B2, B3, and B4. The output of the switching system is connected to digital trunks X and Y, each also serving four digital lines. The four samples carried on trunk A enter TSIU-A in sequence and are stored in sequence in its buffer memory. Similarly, the four samples carried on trunk B enter TSIU-B and are stored in sequence in its buffer memory. The four samples stored in TSIU-A and TSIU-B are read out in a different order over output lines. Each line connects to the time-multiplexed switch (TMS), which is a 2-by-2 matrix switch. The samples are entered into TSIU-X and TSIU-Y in an order necessary to accomplish the switching.

The TMS reconfigures itself four times in $\frac{1}{8000}$ second. Consider the first time interval. Sample A2 from TSIU-A and sample B2 from TSIU-B are switched by the TMS so that A2 goes to the second memory location in TSIU-Y's buffer and B2 goes to the fourth location in TSIU-X. The samples in each output TSIU are read out

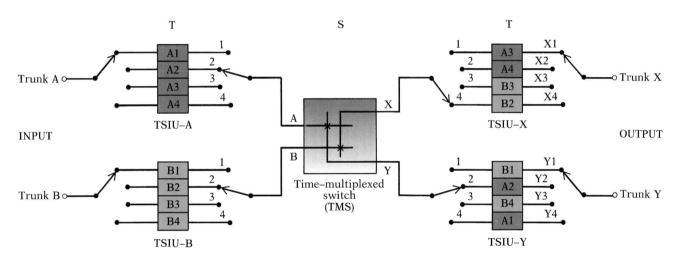

Time	TSIU-A	TSIU-B	TMS		TSIU-X	TSIU-Y
1	2	2	A–Y	B–X	4	2
2	1	3	A–Y	B–X	3	4
3	3	1	A–X	B–Y	1	1
4	4	4	A–X	B–Y	2	3

A digital switching system combines time and space switching.

in order such that A3 is connected to output line X1, B1 to Y1, and so on, thereby accomplishing the following switching:

$$A2 \rightarrow Y2 \qquad B2 \rightarrow X4$$
$$A1 \rightarrow Y4 \qquad B3 \rightarrow X3$$
$$A3 \rightarrow X1 \qquad B1 \rightarrow Y1$$
$$A4 \rightarrow X2 \qquad B4 \rightarrow Y3$$

This type of switching machine performs time-space-time switching and would be referred to as a TST switch.

Examples of Time-Division PCM Switching Systems

The first commercially successful time-division PCM switching system, the AT&T No.4 ESS™ toll switching system, was put in service in 1976. It is used mainly for the switching of circuits in the long-distance network or for a large tandem office. All switches in the No.4 ESS are electronic. The No.4 ESS can process over one-half million telephone calls during a peak busy hour.

The No.4 ESS switching system contains time-slot interchange units at its input lines and output lines, sandwiching a time-multiplexed switch. Seven digital trunks enter each of the 128 input TSIUs. Each trunk carries 128 time slots serving 120

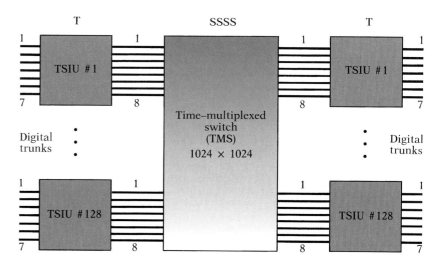

The No.4 ESS™ switching system.

digital voice lines, with eight slots as spares and for maintenance. The output from each input TSIU is carried on eight digital trunks on the time-multiplexed switch (TMS). The TMS can switch 1024 input trunks to 1024 output trunks. Each input trunk to the TMS carries 128 time slots, although only 105 are occupied on the average, and hence the TMS must reconfigure itself 128 times in $\frac{1}{8000}$ second, or 1,024,000 times per second. The output trunks from the TMS connect to another 128 TSIUs. Each TSIU serves 840 digital voice lines, and since there are 128 TSIUs, the total capacity of the No.4 is 107,520 one-way voice lines, or 53,760 two-way circuits.

AT&T's 5ESS® PCM switching system was first installed in 1982. It is a general-purpose switching system used in both long-distance and local switching. It can serve up to 110,000 subscriber lines and can process 300,000 calls per hour. The No.5 ESS consists of as many as 190 time-slot-interchange switching modules.

The 5ESS® time-division PCM switching system consists of as many as 190 switching modules (SM), each under the control of its own processor. Each SM is connected by two pairs of optical fibers to a central time-multiplexed space-division switch located in the communications module (CM). An administrative module performs such common centralized functions as billing, trunk routing, and translation.

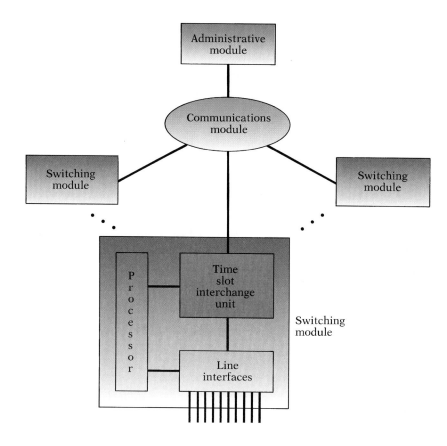

Each switching module is connected by two pairs of optical fibers to a time-multiplex space switch in a single unit called a communications module. One fiber strand carries 32,768,000 bits per second.

Each switching module has its own processor and memory. A single administrative module handles such centralized processes as billing, trunk routing, and translation. The No.5 ESS is truly a distributed control PCM switching system, somewhat akin to the step-by-step switching system except that the flexibility and intelligence of stored-program control is used in each switching module.

A number of other manufacturers also supply digital switching systems. One of the earliest such manufacturers was Northern Telecom with their DMS™ family of central office and toll switching systems. Ericcson, Alcatel, Siemens, Fujitsu, and NEC also manufacture such systems.

Network Switching and Routing

From each phone jack on the walls of your home or office, a pair of copper wires goes to a cable where it joins a collection of pairs of wires, called cable pairs, which form subscriber loops. All the individual wire pairs in the cable can be conceptualized as a form of space-division multiplexing, since each wire pair carries a separate voice circuit. In a large building, a large number of pairs will go to a terminal strip somewhere in the building; there they can be connected to a cable leaving the building.

The subscriber loops, or, more simply, loops, connect a customer's premises with a wire center, usually a local switching office called the central office. It would be wrong, however, to think of a loop as always being a particular pair of wires in a particular cable. As customers are dropped or added, cable pairs are continually reconnected in boxes located in manholes, on poles, or by the side of city streets. Loops terminate in a wire center on a main distributing frame, or mainframe. Thence they are connected to the switching system. Many businesses have their own private switching system, called a private branch exchange (PBX), usually located at the business and used for internal communication as well as for interfacing with the public telephone network. By owning or renting their own switching system, many businesses believe they save money and gain greater control over their internal communication. Instead of having its own switching system, a business can always obtain service from the switching system at the telephone company's local office.

Whatever form of switching is used—manual or automatic, space-division or time-division—many calls must go through several switching offices before they are completed. In national or worldwide telephone systems, your call goes first to an electromechanical or electronic switching system in a local office. If you are calling a party who is connected to the same local office, the local office can complete your call. If not, your call will be sent to another local office or to a tandem office, which interconnects several local offices. If you make a long-distance call, the local office sends your call over a trunk to a long-distance (or toll) switching system. Then the call goes over a long-distance trunk, or circuit, to some distant toll switching system and from there to a distant local office.

Trunks are the cable pairs or other circuits that interconnect wire centers. Trunk circuits may consist of a large number of individual voice circuits that are all combined to share a single trans-

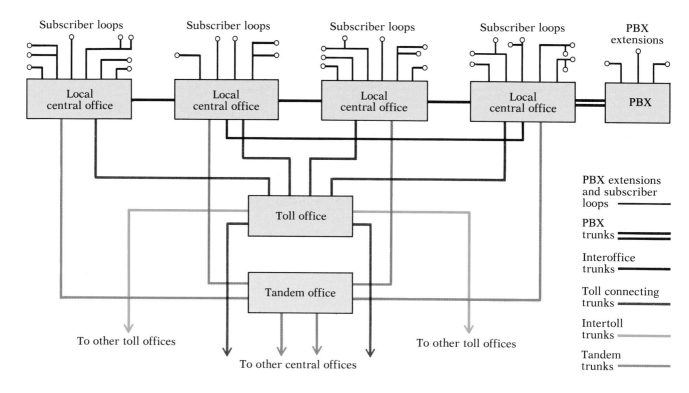

The telephone switching system connects local calls through local offices, and routes long-distance calls through toll offices, which forward calls to distant local offices.

mission path by means of time-division multiplexing using the T1 system. If we make up a long voice-frequency circuit for either voice or data, it is unlikely to be a pair of copper wires from end to end.

Blocking occurs throughout a telephone network because only a limited number of trunks connect telephone offices. If a call is blocked along one path, a telephone switching office tries to find an indirect route through several other telephone offices.

What if the number of call attempts suddenly increases? A large number of long-distance calls are attempted during some holiday periods, because of good will and bargain prices. During such a period of heavy load, alternate routing may actually decrease the chance of completing a call, because it creates a large number of long, indirect circuits, which tie up too many trunks as

CHAPTER 8

they zigzag between cities. Hence alternate routing is decreased or abandoned when traffic becomes very heavy. When a disaster or other emergency occurs, a flood of incoming calls from family members and friends can paralyze telephone service within or out of the area. Incoming calls must be refused in order to restore service.

There is a very good reason why the first commercial time-division switching system, No.4 ESS, handles toll calls rather than local traffic. We noted in Chapter 5 that all long-distance transmission uses four wires for each call—that is, separate paths are provided for each direction. A toll switching system is necessarily a four-wire system, because it must make connections over long-distance circuits. Today many, if not most, four-wire circuits are time-division PCM circuits. Thus everything favored toll switching as the field for time-division PCM switching, and No.4 ESS is a great success. Its capacity of 107,520 trunks is about four times the capacity of the preceding electromechanical crossbar switching system for long-distance circuits.

The Merging of Switching and Transmission

Traditionally, telephone transmission and telephone switching have been kept separate. Transmission has handled weak signals produced by distant telephones. Switching has routed strong signals produced by nearby telephones. But when time-division PCM switching is used in connection with time-division PCM multiplex transmission, this distinction between switching and transmission disappears. The signals are essentially the same for both functions. In both switching and transmission, separate paths are provided for each direction in the conversation. With the addition of a little logic and memory, a time-division multiplex terminal can rearrange the time order in which the digital signals are sent out, and so it can become part of a switching system.

The Era of Digital

In the long run, we should look forward to a mainly digital telephone system that mixes local and toll multiplexing and switching functions in novel ways. The toll network is now almost entirely digital. Optical fiber is the transmission medium of choice. Microwave and coaxial cable transmission systems already in place are being converted to digital, and whatever analog multiplexing equipment exists is quickly being updated to time-division PCM multiplexing. Most switching systems in the toll network are already digital machines.

Except in remote areas of most industrialized nations, local switching is virtually all electronic, and nearly all new installations use digital switching machines. In the mid-1990s, the older-generation, analog, stored-program control, electronic switching systems, such as the No.1 ESS, will be replaced with time-division PCM switching systems. Interoffice trunks are already nearly all digital.

What is not digital is the local loop and the telephone instruments in our homes. The nature and pace of evolution toward all-digital systems at the local level is not clear. In the future, it may become practical and economical to convert between analog and PCM at the telephone handset or the subscriber's terminal. Large-scale integrated circuits, new sorts of PCM encoders and decoders, electret microphones and transistor amplifiers in place of carbon transmitters, and tone ringers in place of bells make conversion feasible. But the telephone company will have to replace or otherwise manage all the twisted pairs of copper wire returning to the central office. It might make sense to time-division multiplex many of them together in the field and carry the multiplexed circuits back to the central office over optical fiber for direct connection to a digital switching system. It might even be practical some day to bring fiber directly to each home.

Boolean Algebra and Digital Logic

For the designers of digital devices, the concepts of an area of very pure mathematics—Boolean algebra—are as powerful as the ideas of sine waves and complex numbers are to those who deal with linear circuits. In digital circuits, there are two possibilities: the switches are either open or closed, and circuits are simply on or off. Boolean algebra is a branch of mathematics that likewise deals with only two alternatives. Bertrand Russell wrote of it in 1901 in an essay entitled "Mathematics and Metaphysics": "Pure mathematics was discovered by Boole, in a work he called *The Laws of Thought*." Indeed, in 1854 George Boole did publish a book entitled in part *An Investigation into the Laws of Thought*. But was this mathematics in fact pure mathematics? A friend of mine once proposed a test that he now refuses to acknowledge. "Pure science," he

Boolean algebra—the mathematical foundation of the workings of digital machines—is named after its founder, George Boole. In 1854, Boole made public his algebra for dealing with the truth and falsity of logical propositions.

said, "is science you can prove isn't good for anything." By this test, Boole's work cannot be pure mathematics. Claude Shannon demonstrated that this is so in 1938 in his master's thesis, "A Symbolic Analysis of Relay and Switching Circuits."

In those early days, transistors were not known, and so the practical implementation of the mathematics was in terms of relays. A relay is an electromechanical device consisting of two electric circuits. An input circuit includes a coil of wire. Current flows in that coil, creating a magnetic field which attracts one end of a hinged piece of metal, called the armature. The movement of the armature causes the closure (or possibly the opening) of electric contacts in an output circuit.

Boole dealt with the truth or the falsity of logical propositions. He represented truth and falsity using numbers, 1 for truth and 0 for falsity. Because propositions are either true or false, only two numbers, 1 and 0, are needed or allowed. Shannon needed just the same two numbers to deal with digital circuits: 1 can represent a current through a relay coil or the closure of a relay circuit (which can cause current to flow through another coil); 0 can represent no current or an open contact.

In the circuits shown in the margin, A and B represent the states or values of two switches. When switch A is open, $A = 0$; when switch A is closed, $A = 1$. If two switches are in series (top), the circuit they form is described by Boolean multiplication as AB (read as A and B, not as A multiplied by B). The product is 0 unless both A and B are 1. Therefore, if either switch is open (0), the circuit is open. If two switches are in parallel (bottom), the circuit is described by Boolean addition as $A + B$ (read A or B, not A plus B). The sum is 1 unless both A and B are 0. Therefore, if either switch is closed (1), the circuit is closed. Boolean variables can be designated by A, B, C, and so on, each variable having a value of 1 or 0.

Besides the idea of the variable representing an open (0) or closed (1) switch or relay contact, we need the idea of the complement of a variable. The complement of A is written $\bar{A}$. If $A = 1$, $\bar{A} = 0$; if $A = 0$, $\bar{A} = 1$. The complement is important since the motion of the armature of a relay can cause some contacts to close (1) and others to open (0).

By using Boolean operations to describe the interconnection of relays, Shannon found a way to express mathematically the behavior of complicated relay networks in terms of functions of several Boolean variables. One set of variables represents the inputs, each of which must be on (1), or off (0). The other set of variables represents the outputs, each of which must be contact closed (1) or con-

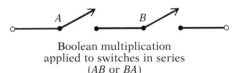

Boolean multiplication
applied to switches in series
(AB or BA)

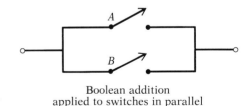

Boolean addition
applied to switches in parallel
($A + B$ or $B + A$)

Switches are placed in series or in parallel, depending on whether two states are to be added or multiplied.

Relay network	Boolean function
A B	AB
A B	$A\bar{B}$
A / B	$A + B$
A / B	$\bar{A} + \bar{B}$

The design of the relay network determines the output of the system.

tact open (0). A Boolean algebraic expression represents the relation of the outputs to the inputs.

By applying Boolean algebra to relay circuits, engineers can design a logical network for which the outputs are related to the inputs by any specified Boolean function. More than one network will give the same Boolean function, although the function will appear in different forms, and by the use of various theorems the form of a Boolean function can be changed so that it corresponds to a simple logical network of relays that will implement the specified Boolean function. These Boolean functions can perform various arithmetic operations, such as addition and subtraction, and logical operations, such as comparing one number to another. They are the basis for the operation of modern digital computers and apply to the solid-state logic of integrated circuits used in computers.

Signaling: How We Operate It All

Another essential function, digital in nature, pervades even the analog portion of the present telephone plant. In a telephone network, we must transmit more than the human voice. We must transmit instructions for operating switching mechanisms.

When a subscriber lifts the telephone handset from its cradle, the hookswitch closes and electric current supplied from the central office flows through the telephone set. This flow of current alerts the central office that service is desired. The central office returns dial tone when it is ready to receive the telephone number of the party being called.

When the subscriber dials an old-fashioned dial phone, the dial clicks interrupt the flow of current and signal the central office. The dialing of a 3, for example, would cause three pulses, or interruptions, in the flow of current. Dial pulses are slow, and since direct current is not sent through the switching equipment, dial pulses stop at the first stage of switching. Dialing by the newer touchtone method produces for each button pressed a unique combination of two single-frequency tones, or sine waves. The tones travel down the line to the central office, where filters detect the different frequencies and in this way decode the dialed digit. Touchtone dialing is fast, and, unlike dial pulses, touchtone signals can be sent over the talking path of the network.

When anything other than a simple pair of metallic wires is used for transmission, some means for signaling must be provided.

The phone companies of the world are converting to common-channel signaling (CCS), previously called common-channel inter-office signaling (CCIS) and today also called signaling system 7 (SS-7), in which the necessary signals go over circuits that are quite distinct from the talking path. CCS speeds setting calls up, because signaling need not wait until the talking path is set up. With CCS, the telephone number of the calling party can be passed along to the local office that serves the called party. This allows a host of new call management services, such as the ability to know the phone number of the caller before answering the telephone. Also, since the CCS signals go over a separate data network, network signaling is secure.

CCS enables the switching machines that operate the network to exchange information. The network can thus be dynamically reconfigured depending upon traffic patterns that change during the day.

The switching machines in the network determine the optimum path needed to provide the voice connection between the calling and called parties. Ultimately, the ringing signal is transmitted over the local loop to the called party's telephone set, which then rings. If the called party is at home and wishes to answer the ringing phone, the party lifts the handset, electric current flows, and the local office knows that the party has answered the call. Ringing ceases and the final step in establishing a voice connection between the two parties has occurred. Conversation begins with the familiar "Hello."

Signaling is, of course, only one aspect of the whole process of switching. It is by means of switching that a vast, worldwide communication network is continually reconfigured to enable pairs of subscribers to talk to each other.

The Functional Network

The programmable computers that control switching machines have given us a network capable of many new and wonderful services. Yet many of these services are simply new incarnations of services of the past when human operators controlled switching.

The human operators of Alexander Graham Bell's day and later provided many "intelligent" services. If I was not at my desk, the operator could take a message or could forward the call to where I might be. If I did not wish to be disturbed, the operator could be instructed not to ring my phone unless my boss called. If

I went to lunch at the corner coffee shop, the operator would know where I was and could reach me in a dire emergency. The intelligence of the human operators gave the network of the distant past considerable functionality.

Human operators were so costly that few could afford telephone service. To bring affordable, universal service, we automated telephone service using electromechanical switching machines. But the electromechanical switching machines were stupid and inflexible, and we lost the personal services that the human operators provided.

In contrast, the new computer-controlled electronic switching systems are programmable, flexible, and intelligent. Functionality thus has returned with such services as call forwarding, call waiting, voice mail, and speed dialing. All these "new" intelligent services simply mimic what was possible a long time ago with human operators. New technology makes these services affordable and universal for all.

The intelligent network continues to evolve, finding new ways for computer-controlled switching systems to communicate over the common-channel signalling (CCS) network. One exciting service that is available in some regions of the country tells you who is calling before you pick up the telephone. The telephone number of the calling party is passed along the CCS network and then sent down the telephone line between the first and second ring. Special equipment at your telephone decodes and displays the number of the calling party. You can also instruct the switching machine to block certain numbers, perhaps from crank callers or bill collectors. The same screening function could be achieved with a device on your line that receives the telephone number of the calling party and allows your phone to ring only for preprogrammed numbers. User-owned device or intelligent switching system? This is one of many decisions to be made about whether the intelligence should be in the network, the terminal, or both.

The processing power and intelligence located in the network makes it possible to record messages there rather than in telephone answering machines at your home. You could leave a message for someone even if he or she did not have an answering machine. When the person first picked up their phone, a special tone would announce that your message was waiting. But if you had a machine at home that could keep trying the number every hour or so until the party answered, the network recording service would not be needed.

In AT&T's 5ESS® switching system, small electronic circuits are joined to form larger circuits. Left: Integrated-circuit chips are interconnected by printed gold wiring on a hybrid integrated circuit. Middle: Hybrid integrated circuits along with other components are interconnected by printed copper wiring on an epoxy board. Right: The boards are plugged into equipment bays, seen here from the rear, and are interconnected by conventional copper wire. Copper wire, or in some cases optical fiber, interconnects bays.

Telephone companies favor placing as much intelligence as possible in the network, but equipment manufacturers favor placing the intelligence at the terminal. Clearly, each party has a vested interest in its own approach. It seems sensible to place the intelligence for customized services as close as possible to the user. After all, most computer users work on personal computers rather than a centrally located computer that a large number of users share. However, communication is not a stand-alone affair, and communication terminals are useful only in conjunction with a network to interconnect them. I therefore believe that intelligence in the terminal makes sense for many customized services, but that appropriate intelligence in the network to facilitate switching and transmission is necessary too.

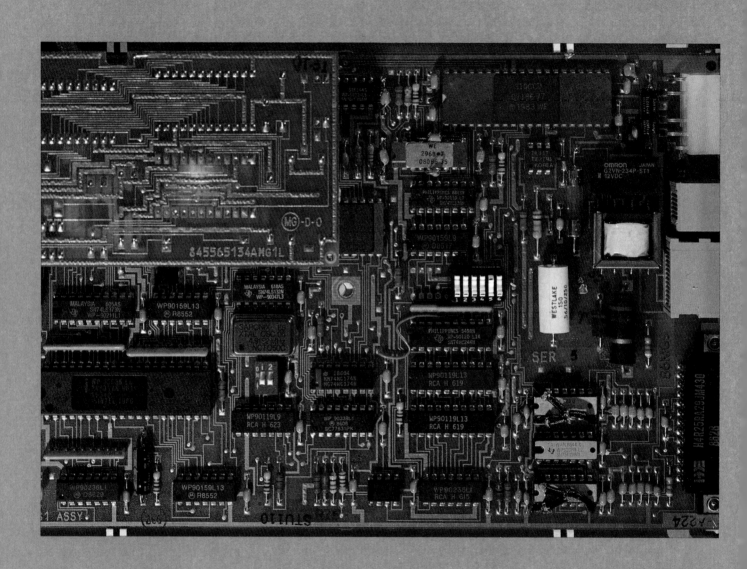

9

$—--.$

Networks for Voice and Data Communication

The dots and dashes of telegraphy were the very first electric signals used for communication. Morse code represented each letter by a set of dashes and dots—a form of binary code. In today's data communication, we have expanded upon Morse's idea, and each alphanumeric character or symbol is represented by a binary code of a single byte consisting of 8 bits. Computers process data encoded in this form, and data communication is a way for computers to communicate with each other directly.

All electronic communication—broadcast radio and TV, telephony, data—is implemented by a collection of interconnected equipment, called a communication network. TV networks carry video and audio signals. Radio networks carry audio signals. The telephone network carries mostly voice signals and facsimile im-

This circuit pack is part of an AT&T modem. Used in data networks, it allows transmission of digital computer signals over analog voice circuits at speeds up to 2400 bits per second.

ages. Data communication uses the telephone network, as well as specialized networks, but it has its own peculiarities.

Facility and Service Networks

We can think of networks in two different ways: as facility networks or as service networks. A facility network is the collection of transmission paths, switching equipment, and signaling facilities

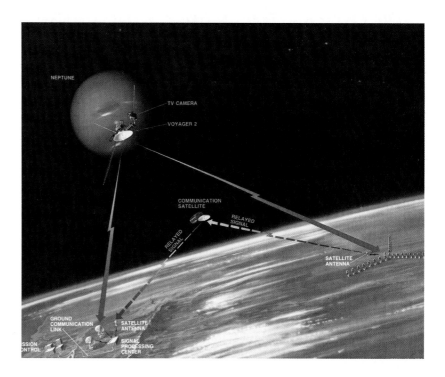

The Deep Space Network, operated by the Jet Propulsion Laboratory for NASA, transmits signals from around the world to and from spacecraft. As Voyager *flew past Neptune in 1989, it sent signals to arrays of antennas spread over the earth, whence the signals were transmitted to signal-processing centers and transformed into a standard brief form. Satellites and ground channels sent the signals to the mission control center, where they were converted into images of Neptune almost 3 billion miles away.*

CHAPTER 9

necessary to provide communication. A service network makes use of various facilities or equipment to provide a particular type of service, such as telephone service.

The interconnected facilities of many telephone companies and long-distance carriers provide one world-wide service network that gives us both local and long-distance telephone and facsimile service. The facilities of private networks—switched and unswitched, civilian and military—can provide telephony; airline reservations; credit-card verification; remote computing for universities, industry, and government; and the distribution of television and radio programs.

Simple networks have evolved into complex networks. Thus, the internal microwave-communication network of the Southern Pacific Railroad became SP Communications, a common-carrier communication company, which later became US Sprint; under that name it provides long-distance service throughout the United States in competition with AT&T and other companies. Transmission is cheapest and most saleable when the facilities are fully loaded with all sorts of traffic and when a wide geographical area is served.

This communication satellite receiver on an oil platform in the North Sea is part of a private industry network.

The Challenges of Data Communication

Many people compose reports, draft letters, even write novels on their personal computers. The report or letter in the computer can be transmitted as digital data to another computer or to an electronic storage medium, called a data base, located far away. Similarly, text stored in a data base can be transmitted to a personal computer. Most text goes to and from personal computers over the public telephone network.

Now that so many of us have computers in our homes and offices, we should be able to zip a message to its destination by pressing a few keys. Why do not many more of us send "electronic mail" from home to home or office to office? The lowest night telephone rate from New York City to Los Angeles is eleven cents per minute, and rates are continuing to decrease. In one minute, we could certainly send 3000 or perhaps even 6000 words (10 or 20 double-spaced typewritten pages at 2400 or 4800 bits per second). This would be cheaper and faster than mail.

The would-be sender of electronic mail needs a personal computer with keyboard and display. In addition, a device called a

Electronic Mail and Facsimile Transmission

Today there are two competing ways of transmitting text: electronic mail, in which words are spelled out character by character, and facsimile transmission, or fax, in which a black-and-white reproduction of a page of text is transmitted.

Electronic mail transmits the bare essentials of text, retaining characters and numbers, upper and lower case, and various useful symbols by using 8 bits per symbol. Single-spaced typewritten text has about 5000 symbols per page. At 8 bits per symbol, this is 40,000 bits per page.

The picture received in fax retains type style, layout—indeed, everything but color and shading. The received picture is made up of about 2200 lines of about 1700 picture elements each, for an $8\frac{1}{2}$ inch by 11 inch page at a resolution of 200 lines per inch. These picture elements could be transmitted simply by using 1 bit (denoting black or white) per picture element, or $2200 \times 1700 = 3,740,000$ bits per page.

A 1934 telephotography machine was bulky and difficult to operate.

Actually, documents transmitted by fax tend to have long stretches of white and/or black. By specifying the length of long stretches of white and/or black rather than by sending the same code for each point, transmission time can be greatly decreased.

modem is required to enable the computer to send and receive data over the telephone network. The costs are substantial, but the greatest problem for data communication is the lack of standardization, both of transmission standards and of protocols for handling calls. Facsimile terminals are nearly as expensive, yet facsimile is spreading like wildfire. The secret is standardization.

Such an ingenious "run length" code has been standardized world-wide for fax transmission, and is built into all office and home fax machines. This code cuts down the transmission time for a page of single-spaced text by a factor of about 9, so that only about 400,000 bits are needed to transmit such a page by fax. This is, however, nearly 10 times the number of bits required in electronic mail.

Currently, fax transmission over ordinary telephone lines is growing faster than electronic mail. Fax can transmit letters in nonalphabetic languages such as Japanese, and fax is widely used in Japan. Fax will transmit any black-and-white page, including letter-heads, penciled comments, and sketches and longhand. Thus, it is more powerful than electronic mail, and in many ways more convenient. A typical fax transmission over a telephone line takes about a minute per page; that isn't bad. The faster transmission of electronic mail may not offer a substantial advantage.

Modern fax equipment is compact and easy to use.

Perhaps most important, fax is standardized worldwide. Over a telephone connection, any equipment will operate with any equipment, while the various sorts of electronic mail systems may interconnect or not. Standardization is a powerful advantage. What businessperson today does not have a fax number on the business card?

Protocols

When one person speaks to another, a set of rules and procedures governs the conduct of the conversation. The person answering the telephone says, "Hello," and the calling party says, "Hello, this is John calling." We know it is rude to speak while the other person is speaking, and we await a pause in the other person's speech before interrupting. For long stretches of monologue, the listening party grunts an "ah ha" or "hmm" so that the speaking party knows the listener is still patiently and attentively listening. If something is not understood, we know to ask the other person to repeat what was said. At the end of the conversation, we each say, "Goodbye." The set of rules governing the conversation are called protocols.

Computer or data terminals must follow rigid, unambiguous protocols to communicate with one another. These data protocols have been likened to handshakes. The called machine must answer the call and tell the other machine when it is ready to receive data. If one machine detects an error in transmission, it must request the other machine to retransmit the data. The two machines must terminate the communication in an orderly fashion. These protocols must be agreed upon by both machines.

Users often cannot send data from one machine to another because different types of data bases and computers follow different protocols. One solution is to add a place or gateway in the network where conversion from one protocol to another can take place, so that many different protocols can all appear to be the same to a user or computer. Such gateways can interconnect different data networks, data bases, or computers.

The International Standards Organization has adopted a model for data communication called Open System Interconnection (OSI). The model defines seven layers for which protocols are needed for communication across locations within a layer. The highest layer concerns the specific application, for example, database access or electronic mail. The lowest layer concerns the physical connection of the modem to the computer.

Standards

In voice communication by telephone, certain standards have been agreed upon: the size of the handset, the type of modular plug leading from phone to loop, the touch tone frequencies, among others. Similar standards are needed to govern data communication.

For simple data communication over the public switched network, a modem between the computer and the network must convert the binary data into the modulated tones that are carried over the network. A similar modem is needed at the other end of the connection to demodulate the received tones. For the return transmission, the role of each modem reverses; the modem is a *modulator-demodulator*. The modem's carrier frequencies, method of modulation, and rate of transmission must be standardized. The protocols for communication are embodied in the computer's software, not the modem, but they too must be standardized.

Data standards are needed for more elaborate data networks utilizing message and packet switching, described later in this chapter.

Transmission of Data

The digital data must modulate tones (sine waves) that can be transmitted over the telephone network as carrier waves. The tones vary in maximum amplitude, frequency, or phase, depending upon the standards used for the particular speed of transmission.

In the simplest form of data modulation, the carrier sine wave is turned on and off, a form of modulation called on-off keying, or OOK for short. The term *keying* finds its genesis in the early days of telegraphy when the telegraph key turned an electric current on and off. Alternatively, the amplitude can be shifted from one level to another, a form of modulation called amplitude-shift keying, or ASK for short. OOK is a special case of amplitude-shift keying. The frequency of the carrier sine wave can be changed from one frequency to another, a method called frequency shift keying, or FSK. Or the phase can be one value or another, a method called phase shift keying, or PSK. Since absolute phase is difficult to determine, the phase usually is shifted from one value to another and the difference in phase is measured, a method called differential phase-shift keying, or DPSK. And a combination of these methods can give faster speeds of transmission.

The bits of a character can be transmitted over a telephone connection at various standard rates, including 300, 1200, 2400, 4800, 9600, and 19, 200 bits per second. The method of modulation is different for different rates. For the 1200 and 2400 bits-per-second rates, DPSK is used. Successive 1800-Hz signal "pulses," known as bauds, are sent at half the bit rate. As shown in the figure on this page, the phase of each pulse differs from the phase of the last pulse by one of four angles, 45, 135, 225, or 315 degrees. This gives four possible patterns of 2 bits per pulse, 00, 01, 11, or 10. Phase-shift keying is also used for the 4800 bits-per-second rate, but with eight different phases instead of four.

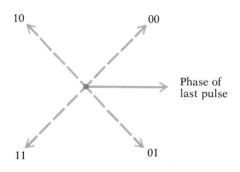

In DPSK the phase of a pulse changes relative to the phase of the preceding pulse by one of four angles: 45, 135, 225, or 315 degrees. Each change in phase represents one of four possible two-bit signals.

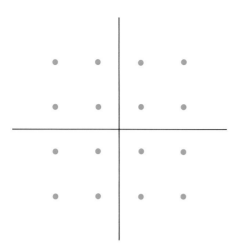

Sixteen possible combinations of phase and amplitude allow the encoding of a pulse carrying four bits.

For the 9600 and 19,200 rates, the successive pulses have different amplitudes as well as different phases. For the 9600 bits-per-second rate, the amplitudes and phases are such that an arrow representing a pulse ends on one of the dots of the 4-by-4 (16 in all) points shown in the figure on this page. Each dot represents a different combination of amplitude and phase. For the 19,200 bits-per-second rate, an 8-by-8 array of 64 points (not shown) is used. This is "more than enough." The extra amplitudes and phases allow the use of error-correcting encoding.

ASCII

Most text is encoded using the American Standard Code for Information Interchange (ASCII). This coding system uses 7 bits to represent various upper and lower case letters, numbers, special control symbols (for example, backspace or carriage return), and other symbols, such as ! @ # $ % & *. The 7 bits are ordered starting with the most significant b_7 down to the least significant b_1. For example, the letter "T" would be represented in ASCII as 1010100. An eighth bit, for error detection, is appended at the beginning of the 7 bits. This error-correction bit, called the parity bit, can be set so that the total number of 1's is an odd number, odd parity. Or it can always be set to 1, mark parity, or to 0, space parity. The terms *mark* and *space* come from the days of telegraphy. ASCII is referred to as an 8-bit code.

Often, personal computers transmit a single ASCII character at a time. Usually, a long series of marks is sent during the idle state of the transmission so that a break in the circuit can be detected by the loss of carrier. However, if the first ASCII bit is also a mark, it cannot be detected without first alerting the receiving computer that the transmission of an ASCII character is about to begin. Hence, the sending computer changes the idle state of the transmission for the first pulse, the so-called start bit. This start bit is then followed by the 7 bits that represent the ASCII character; the least significant bit b_1 arrives first. The last and eighth bit to arrive is the parity bit for error detection. In the example shown on the following page, the ASCII character "T" is followed by an odd parity bit. The signal arrives from left to right. To return the circuit to the idle state at the end of transmission, a so-called stop bit is transmitted. This form of data transmission in which a single character is sent at a time is called asynchronous transmission, as

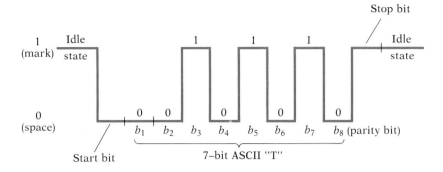

A data signal for transmitting the letter T in ASCII followed by an odd parity bit.

Stop bit

1 (mark) | Idle state

1   1   1   Idle state

0 (space)

0   0   0   0   0

b_1 b_2 b_3 b_4 b_5 b_6 b_7 b_8 (parity bit)

Start bit

7–bit ASCII "T"

opposed to synchronous transmission, in which a whole block of characters is sent in one long transmission.

Some communication, such as broadcast television, is in one direction all the time. Such one-way communication is called simplex communication; it stands in contrast to two-way duplex communication. In full-duplex data communication, both machines can be transmitting and receiving data simultaneously. If a modem is handling full-duplex communication, the same carrier frequencies cannot be used in both directions. The originating modem sends one set of carriers, and the answering modem another.

Data Switching

There are a number of switched data networks—the U.S. government's Autodin network, the Defense Advanced Research Projects Agency (ARPA) network, the airline-reservation networks, various private data networks, and a number of common-carrier data networks that incorporate telephone-company transmission facilities. Some of the switched data networks are akin to the switched telephone network, and, indeed, the switched telephone network is used for data transmission.

When data travels over the telephone network, a continuous circuit is set up between the computers that are communicating. In such a circuit (or line) switched network, the complete connection must be established before data transmission can commence, and the connection is maintained during the data call whether or not any data is being transmitted. Circuit switching is costly for short bursts of data, and it can waste valuable time in setting up a call. However, the public network is ubiquitous and reasonably inexpensive.

NETWORKS FOR VOICE AND DATA COMMUNICATION   205

More elaborate data networks make use of message and packet switching. Either a complete message or a block of data called a packet is deposited in the network and works its way from source to destination. The complete message or block of data is preceded by an address, much like the address on a letter. If the data transmission is not of fixed length, it must close with an end-of-message signal or instruction. The message or packets also contain information about their source. There is no continuing physical connection or circuit from the source to the destination. Connections are made when the message or packet reaches various switching nodes. If necessary, the message or packets are stored at network nodes until an outward circuit is available. A given message or packet finds its way through the network much as a train finds its way to its destination on a network of tracks. Switches are not thrown in advance; they are thrown as the train approaches, and a closely following train may be switched to a different route.

A message is a complete data communication and can be quite long. A packet is a short block of data, usually of a fixed length of about 1000 bits. Individual packets that compose a long message may travel over different routes and arrive at the destination out of order. The computer at the destination must assemble the packets in the correct order.

Data messages are switched from node to node in much the same way that trains are switched from track to track.

ISDN

The switched telephone network provides a uniform grade of voice service between almost all places on the globe. There have been proposals to form an analogous universal switched data network for all, or almost all, data communication.

In telephony, all speakers and hearers can be satisfied by one bandwidth, one signal-to-noise ratio, and the same type of transmitter and receiver (the telephone set). In contrast, the uses of data transmission beyond electronic mail differ from one another profoundly. Airline-reservation and credit-card-verification systems need only slow and simple devices, whereas some businesses transmit voluminous records of each day's transactions at a far higher rate requiring complex high-speed devices and networks.

A telephone customer must be able to reach any other subscriber. In contrast, a credit-card-verification data terminal need be connected to only a few other terminals or computers. Trying to put all kinds of data communication through one universal switched network, comparable to the switched telephone network, may be unnecessarily complicated and costly unless that network is the telephone network itself.

Communication agencies worldwide are peddling ISDN (Integrated Services Digital Network). You can actually have it come to your premises in some lands on a limited basis. It has been standardized internationally in most details—even before coming into existence. ISDN is intended to provide a common digital channel from anywhere to anywhere—the rate starting at 64,000 bits per second, with provisions for going higher. ISDN proposes to deliver to an outlet on your wall two basic two-way digital channels, each at 64,000 bits per second, for basic voice and data service and a single two-way packet-switched digital channel at 16,000 bits per second for signaling. ISDN would thus make possible a true four-wire connection to its network and would also provide a separate signaling path. Japan seems closest to offering ISDN on a widespread commercial basis.

ISDN can provide a wonderful communication channel for voice, fax, or picture transmission, but it poses problems. The subscriber must supply expensive terminal equipment. If only a few people want ISDN, it will not grow rapidly, and the users will be able to reach few destinations. Perhaps its purposes may be better served by digital transmission over the switched telephone network and standardized electronic mail. Someday the telephone network itself seems sure to provide two-way digital transmission

An ISDN system at an Alcatel research laboratory in Italy.

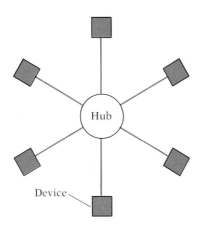

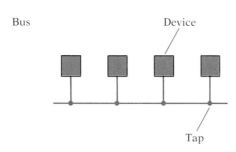

Bus

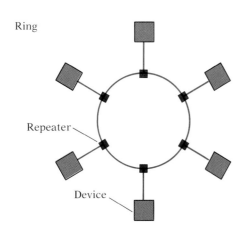

Ring

Star, bus, and ring topologies for local area networks.

right to telephone sets at a rate of 64,000 bits or more per second. Perhaps the switched telephone network will be the universal data service of the future. Or perhaps ISDN will somehow blend with and possibly become the switched telephone network.

Optical fibers can provide digital transmission rates of thousands of millions of bits per second, rates that are far beyond any data-transmission needs that we can forsee. With the exception of mobile service, optical fibers will someday be the universal transmission path in all electronic communication.

Local Area Networks

Many organizations have within a building or adjacent building large numbers of personal computers and other computers, printers, and other devices that must communicate with each other. Such devices can be linked together through a local area data network, commonly called a LAN.

LANs can have one of three different topologies: a star, a bus, or a ring. In a star configuration, all the hardware is connected to a central hub which provides centralized control and switching. In contrast, the bus and ring configurations have no centralized control; control and switching are located at the nodes that join the hardware to the transmission medium.

In the bus and ring, the data signal from one device is received by all the other devices. A data collision could occur if one device were to start transmission during the signal from another device. Rules, or protocols, are needed to avoid such collisions between signals from different devices.

One scheme used to eliminate collisions on a bus configuration is called carrier-sense multiple access with collision detection, or CSMA/CD for short. All devices on the LAN listen continuously to the traffic on the LAN. If a device wishes to transmit but senses that carrier signal is already on the line, the device simply waits. If, however, a device starts transmission and then senses the data carrier of another transmission, both devices will cease transmission and transmit later after waiting different randomly chosen amounts of time. This scheme is used by Xerox's Ethernet™ LAN.

Another collision-avoidance protocol passes a token along from device to device. The token is a unique data code. If a device has data to transmit, it must wait for the token, which is then taken from circulation and returned at the end of transmission. The token protocol is used on both bus and ring topologies. An-

CHAPTER 9

other protocol is the use of a preconfigured data slot, which carries an indication whether the slot is empty or full. A device ready to transmit data must wait for an empty slot, mark it full, and then insert the data for transmission.

Networks Yesterday, Today, and Tomorrow

Today's communication networks are accumulations of facilities built at different times. Most of the facilities were intended for, or grew from, the needs of voice communication. In the switched telephone network, these facilities have provided a revolutionary service that is almost universal in access and worldwide in coverage. What home in most industrialized nations doesn't have a telephone?

Some projections of the future envision an overwhelming volume of data traffic. Yet if you look into the details, it is hard to imagine human beings generating a volume of data traffic in total bits anything like the volume in bits of telephone traffic. If transmission facilities are to be overwhelmed by anything, it is far more likely to be television traffic or picturephone. The table gives data for the United States for 1985 assuming that the listed services are carried totally in digital form. Even if the whole labor force were to work every day of the year at data terminals, telephony would still generate much more digital traffic.

One of the messages of this book is that for a host of reasons all (or almost all) transmission and switching will ultimately be digital, mostly using optical fiber and computerized switching systems. The digital compact disc wiped out the vinyl disc in less than five years. Two-way telecommunication is much more complex as a system. The cost of tearing out the analog part of the telephone plant and, particularly, local loops to replace it with digital transmission and switching facilities would be oppressive and unjustified. We will not have an end-to-end all-digital network for some years.

Yearly Digital Traffic

SERVICE	NUMBER OF BITS
Telephone	7000×10^{15}
Mail (as ASCII)	3×10^{15}
Mail (via Fax)	40×10^{15}
Checks	0.03×10^{15}
Information workplace*	153×10^{15}

* 30 screens per hour at 8 hours per day, 365 days per year, for 114 million workers

POST OFFICE RADIO-TELEPHONE SERVICES

Telecommunication Policy and Challenges

elephone service is universal in the sense that as one can post a letter to almost anywhere, one can reach almost all parts of the world by telephone. A letter may be of no avail if the addressee cannot read; a telephone call may be of no avail if there is no telephone in the neighborhood.

The availability and use of telecommunication in various countries reflects, among other things, their economic progress. We can see this by comparing telephone penetration in some representative countries with gross national product (GNP). In general, telephone penetration is largest in technologically advanced countries with a large affluent society and a high GNP per person, such as Sweden, Switzerland, Denmark, Canada, and the United States.

A 1935 educational wall chart depicts radio-telephone service from London to the rest of the world.

Telephone Administration

The switched telephone network is worldwide. Various national and other telecommunication administrations own and operate the equipment that provides actual service to customers. All are faced with the challenge of providing universal service by means of highly complex technology. How has telecommunication service been administered around the world, and how is it administered?

In discussing this matter, it is helpful and indeed almost essential to understand something of an earlier sort of universal communication service, the postal system. This system, though far less technologically complex and demanding than telephony, exemplifies the same social and economic problems for the supplier of service.

In general, telephone penetration is largest in technologically advanced countries having a large affluent society and a high GNP per person. We by eye divide countries into perhaps three groups: the undeveloped, which have low GNP per person and very few telephones; the developing, which have moderate GNP and moderate numbers of telephone (although Japan and West Germany are in this group); and the technologically advanced nations, which have many telephones and a large GNP per person.

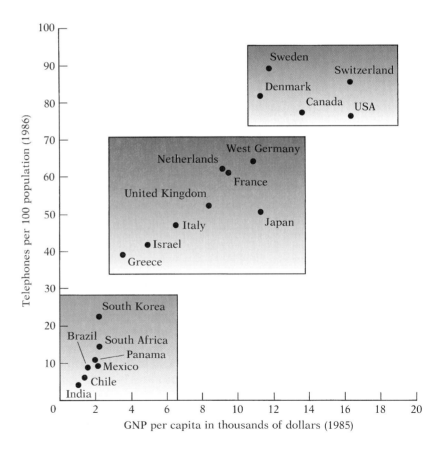

CHAPTER 10

In 1837, a British educator and tax reformer, Rowland Hill, published a book that revolutionized sending and receiving letters, at first nationally, and finally internationally. He set an inescapable pattern for the provision of postal service. Hill's book was titled *Post Office Reform: Its Importance and Practicability*. The revolutionary reforms that Hill proposed, none entirely new, were adopted in 1840. These consisted chiefly of (1) prepayment by affixing a stamp to a letter and (2) a cheap, uniform rate (initially one penny) for a half-ounce letter addressed to *any* part of Britain.

Before the "penny post," charges were based on distance. The smallest charge for a letter was fourpence; the average charge was $6\frac{1}{4}$ pence. Hill's investigations showed that transit expenses were an insignificant part of the total cost of delivering a letter; the irrelevant charging scales based on distance required a host of clerks and inflated the cost of operations.

While the penny post that followed Hill's suggestions succeeded, it could succeed only as a monopoly. Had a competing operation been allowed to provide limited service, such as service in London only, it could have undermined the postal system by what is now called "cream skimming," that is, leaving the provision of costly and unprofitable but necessary service to the government as the supplier of last resort. The conflict between smallest cost to a favored few and smallest overall cost has plagued telephony to this very day.

To be useful to the whole population, a service such as the mails or telephony must be available to all, or to almost all. The post office cannot afford to bypass remote citizens whom it is costly to serve. Neither can telephone companies. Telephony has become so essential a part of our lives that we feel that we are entitled to have telephone service. In the United States, an inexpensive basic telephone service, called Lifeline service, is offered for the poor.

If all, or essentially all, are to be served, the cost of service must be small for everyone, not just those whom it is cheapest to serve. We should note that the "actual" cost of service, if it could ever be determined, would certainly vary widely among subscribers.

Telephone service has a good deal in common with postal service, but it differs from the mails not only in the greater intensity of its technology but in other ways. The nationwide and worldwide transmission facilities of telephony were built *primarily* for telephone traffic, though they serve subsidiary purposes. In contrast, the physical transportation of letters uses a miniscule fraction of the total system of transportation of physical objects. Telephony provides and dominates all electronic message transmission.

The negligible impact of the mails on the world's transportation resources limits the economic influence of even the most successful postal service. But suppose that competing telephone companies were allowed to provide long-distance communication, as is the case in the United States. If a company that provides long-distance circuits for communication is unsuccessful, it has no other sizeable communication market to turn to. Won't the company that provides the cheapest service drive all other common carriers out of the market? Indeed, it would if the government did not intervene. This the United States government does in various ways to maintain some sort of "competition." Many other nations allow no competition at all.

The problems of universal telephone service are like the problems of postal service, only more so. The telephone *is* a universal service in many industrialized nations. The absolute essentials are a reasonable charge for service based on some sort of average cost, avoidance of cream skimming, and a continually advancing technology to support a service that dominates rather than draws on other communication services. And, adequate financial resources to provide service and advance technology are also necessary.

The common recourse in providing both postal and telephone service has been to make them a government monopoly, but the nature of the monopoly has been different in different countries. And, there are alternatives to government monopoly, and they have been changing dramatically over the last few years.

Differing Means for Providing Telephone Service

National and other telephone systems have been operated in a variety of ways, and still are:

- In the past, various national telephone systems have been operated by a private organization such as the International Telephone and Telegraph Company (IT&T), either directly or under contract for the national government. The telephone systems of Spain and of several South American countries were once operated by IT&T. Those systems have now been nationalized as a matter of public policy.

The British Telecom tower, with its many microwave antennas, is a London landmark.

- Some telephone systems are operated as departments of the government, as the United States Postal Service used to be. Both expenses and capital needs are met (or not met) through legislative appropriations.
- A telephone system may be operated as a public corporation, as it was before privatization in Japan and as it is in Britain. Public financing and financing through the sale of bonds support the corporation. Sometimes a separate government-owned corporation handles international traffic, as in Australia and Indonesia.

- A telephone system may be operated by one or more private companies, as it is in the United States and Canada. Private companies are always regulated, and of course pay taxes. In the United States, most telephone and related service was once supplied by the Bell System, which was, in effect, a national system. Today the United States has a national telecommunication service, but no national system; the ownership of the communication network is divided among a number of companies.

Financial Challenges

Whether public or private, a large telecommunication system faces severe financial challenges. Because of taxes, these financial challenges are somewhat greater for private systems than for government systems. In the United States, for instance, state and local taxes are a large item of expense, exceeded only by labor and depreciation. There are also local sales taxes and federal excise taxes that are levied on telephone users.

When a telephone system is operated by a government department or administration, revenues go directly into the treasury. All the money for operating expenses, expansion of plant, and research and development must come through legislative appropriation. Money for telephony commonly comes at the bottom of the list in legislative interest and enthusiasm. Communication is less pressing than defense, welfare, or health care.

The public corporation is an attempt to solve the problem of support. The telephone company is allowed to retain its revenues and may borrow money by selling bonds. Sometimes, however, it is difficult to sell bonds. During the period when Nippon Telegraph & Telephone (NTT) was a government corporation, NTT ingeniously insisted that new subscribers buy bonds to cover the capital cost the company incurred in serving them. In Finland, telephone companies charge subscribers an entrance fee of about $850 for telephone service; payment also gives the member subscriber a single voting share in the nonprofit private companies that provide most local service there. A nonmember rental subscriber is charged a lesser entrance fee of about $300, but must pay about three times as much for annual service.

Private telephone companies retain earnings and sell bonds and stock when they can to raise the funds necessary for expanding and improving the network. In 1987, the Bell operating companies (the local telephone companies that serve individuals directly) spent $14.0 billion in expanding their network. The total revenue of the seven regional holding companies for that year was $70.6 billion. Few other industries, if indeed any, must reinvest at the annual rate of nearly 20 percent of revenues ($14.0 billion of $70.6 billion). In the same year, AT&T spent $2.7 billion of its $17.6 billion of revenues from telecommunication services for network expansion, an annual rate of 15 percent.

Whatever their nature, telephone companies or administrations have pressing financial problems. But adequate money is not a sufficient condition for success. The success or failure of a telephone company or administration lies in its ability to counter through research and technical ingenuity the ballooning capital cost of telephony brought about by increasing volume of service, increasing costs of materials and labor, improvements in the quality of service, and the provision of new communication services.

Research and Development

How do various sorts of telephone systems cope with technological change? Small companies or administrations cannot afford extensive research and development. They buy from various entrepreneurs good, modern equipment that has been developed for a

A scientist at Canada's Bell-Northern Research laboratory investigates a new technique for the fabrication of fiber-optic communication lasers.

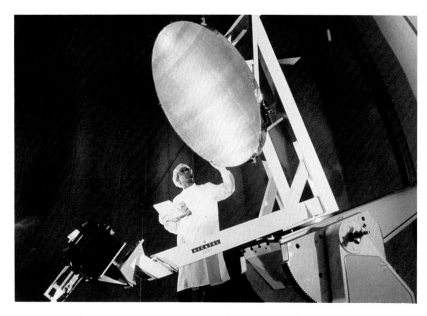

An Alcatel Espace plant in Toulouse, France, tests a reflector antenna for the Eurelsat II *communication satellite.*

larger market. A certain amount of engineering is necessary to fit the equipment to particular needs, but that is far simpler than creating new technology and embodying it in new systems. In essence, small telephone companies or administrations exploit the state of the international communication art without adding to it. That objective can be carried out very well, but some small companies or administrations spend too little on equipment and give poor service.

All national telephone companies or administrations have research and development laboratories of some sort, but nearly all companies must rely largely on outside suppliers. Thus, there is the danger that the telephone administration will not be able to communicate successfully its problems, ideas, and proposals to the actual telephone equipment manufacturers. Further, the suppliers of equipment have research laboratories themselves, and the staff of the telephone administration's laboratories may not have access to the real state of the art as understood by the suppliers.

When research and development are separate from the production of equipment, some way must be found of ensuring cooperation and interchange of information. In many countries, there has been very close long-term collaboration between the telephone administrations and several chosen suppliers. In the United States, this sort of relation might be looked on as collusive collaboration.

The System That Was

For about a hundred years, telephone service in the United States was unique among the telephone systems of the world. It provided the cheapest and best service. Its research and development facilities at Bell Laboratories were rightly regarded as the best in the world. Bell Laboratories developed sound, modern telephone equipment with a strong emphasis on reliability. The system's manufacturing facilities, Western Electric, had high standards of manufacture, or of certification of non-Western equipment.

The Bell System was a vertically integrated monopoly that provided telephone service through the local operating companies and long-distance service through the Long Lines division of AT&T. Its manufacturer, Western Electric, sold its products, including telephone instruments, switching systems, and transmission systems, to the local companies. Central direction was given by AT&T, which also was the ultimate owner of the stock of the various entities.

The pattern and growth of the Bell System was no accident. The vision of Bell himself was not limited to local service. As we noted in the first chapter, Theodore N. Vail set a pattern for the Bell System with his proposal "One policy, one system, universal service."

All manufacturing for the former Bell System was done at Western Electric plants. Here, diecast telephone bases roll off the production line at Western Electric's Hawthorne Works in Chicago.

Vail and the Bell System

Almost as important as the invention of the telephone itself was the invention of the Bell System—an integrated corporate structure for providing telephone service to the public. This structure was the invention of Theodore N. Vail.

Vail began his career with the Bell System as general manager of the Bell Telephone Company in 1878. He later became the first president of the American Telephone & Telegraph Company (AT&T) in 1885; he left AT&T two years later. After pursuing other interests for twenty years, Vail returned as president of AT&T in 1907, retiring in 1919 as Chairman of the Board. Vail believed in "One policy, one system, universal service." He regarded telephony as a natural monopoly, that is, a service that is cheapest to provide and best

if there are no competing systems and providers. He saw the necessity of regulation and welcomed it. Vail was a true visionary who developed and emphasized the long-term objectives of the Bell System over short-term financial objectives. During his tenure from 1907 to 1919, earnings per share actually drifted downward somewhat. At the right Vail is pictured six months before his death in 1920.

Vail believed in the importance of a central research and development laboratory. It was Vail and his belief in the consolidation of R&D that created the foundations for Bell Labs as the central R&D unit for the Bell System. The results of this research were developed into products and services, also designed at Bell Labs to ensure tight coupling between research

and development. The products of development work at Bell Labs were manufactured by Western Electric, acquired by Vail in 1881 as the exclusive manufacturing arm of the Bell System.

Economical, universal, *good* service was sought through science, technology, good business practices, and eternal vigilance. Bad financial returns could send a shudder through the Bell System. So could poor service indices (the telephone operating companies, now the Bell companies, had to keep detailed records of quality of service, including delays and wrong numbers).

The research at Bell Laboratories was, all in all, superb. It was suited, not to a manufacturing company whose revenues fluctuate wildly with the economic cycle, but to the future needs and opportunities of a national resource that was there yesterday, was there today, and was expected to be there tomorrow and tomorrow. Research was accordingly supported largely by a fee levied by AT&T on the operating telephone companies.

Divestiture

For many years, the Federal government attempted in some way or another to break up the Bell System. The System survived, through shrewd compromises over the years, by disposing of its ownership of the telegraph company Western Union (in 1914), by selling its radio network (in 1926), and by licensing all patents and restricting itself to telecommunications (in 1956). When, in 1974, the United States Department of Justice launched another antitrust case against the Bell System, there was nothing extraneous to give up.

AT&T's management, the Department of Justice, and the court could have chosen to follow a pattern more like that of telephone administrations abroad by divesting AT&T of its manufacturing facilities. The country would have retained a national telecommunication system, with a Bell Labs that did research and advanced development only, and no manufacturing would have been allowed within the system. Instead, on January 1, 1984, AT&T divested itself of the Bell operating companies, whose stock was given to seven newly created regional holding companies, all independent of AT&T and of one another. AT&T continued to supply long-distance service to the customers of the Bell companies, in competition with other long-distance companies. And AT&T retained both Western Electric and AT&T Bell Laboratories, with the freedom to go into all fields of manufacturing, including computers, which had been forbidden territory before the breakup.

Afterward

When the Bell operating companies left AT&T, part of Bell Laboratories went with them as Bell Communications Research (Bellcore), a research organization to serve the divested Bell companies jointly. Those companies, though forbidden to manufacture, nonetheless are developing small independent research organizations of their own in addition to relying on Bellcore.

What has happened to Bell Laboratories? A few years ago one of the authors (Michael Noll) was asked by the National Science Foundation to study the effects of divestiture on Bell R&D. I examined budgets and number of employees, patents and publications. I also spoke with a number of my friends and colleagues who

Contributions of Bell Scientists to the Art of Communications

Work at Bell Laboratories, and elsewhere in the Bell System before the Laboratories were founded in 1925, has abounded in discoveries, inventions, and advances in the art of communications.

In the field of transmission, Harry De Forest Arnold understood Lee de Forest's primitive audion and turned it into a useful vacuum-tube amplifier for telephone repeaters. In 1915 George Ashley Campbell invented the electric-wave filter, a device essential to frequency-division multiplex transmission, and in the same year John R. Carson invented single-sideband transmission, which is used in all frequency-division multiplexing. The negative-feedback amplifier, invented by Harold S. Black in 1927, made frequency-division multiplex transmission stable and economical. Harry Nyquist devised the Nyquist criterion for the stability of negative-feedback amplifiers,

and Hendrik W. Bode discovered a relation between phase and gain and showed how Nyquist's criterion could best be met.

In 1947 Walter H. Brattain, John Bardeen, and William Shockley invented the transistor, which greatly expanded the capabilities of transmission and which ushered in a totally new era of electronics and computer technology. In 1948

Claude E. Shannon summarized the entire process of communication in his information theory.

The first transcontinental microwave telephone transmission system grew directly from the work of Harold T. Friss and his colleagues at the Holmdel, New Jersey, laboratory. Many of these same people were involved deeply in work on the *Echo* communication

worked at AT&T Bell Labs and at Bellcore. In terms of funding and people, both AT&T and the phone companies continued their commitment to R&D after divestiture. Nor did there appear to be a slackening in publications or patents at AT&T Bell Labs.

Still, AT&T Bell Labs is not the same as the old Bell Labs that I worked for. Some areas of research have been de-emphasized or abolished, for example, psychology and economics. Bell Labs was a great research institution with a clear mission; it was responsible for the future of telecommunication in the United States. It had secure long-term funding provided by the stable Bell System income from telephone operations. All that has changed with divestiture. AT&T appears to be unsure where its future lies, though there satellite, which carried voice signals across the continent in 1960, and in work on the *Telstar* satellite, which first carried telephone conversations and television programs across the Atlantic in 1962.

The Bell System has pioneered in switching as well. Edward C. Molina applied the theory of probability to blocking in switching systems early in the century. He also invented translation in switching, a process that allows the free assignment of telephone numbers to subscribers. Raymond W. Ketchledge produced in No. 1 ESS the first commercial electronically controlled switching system, and Earle Vaughan developed the first commercial time-division PCM switching system, an outgrowth of work he had done years earlier in the research department.

Work directly applicable to telephony by Harvey Fletcher and his colleagues, whose tests set standards for telephone transmission, led to the first sound motion pictures and the first truly high-fidelity reproduction of sound. Fletcher and his colleagues also developed the audiometer and conducted the first large-scale tests of hearing at the New York World's Fair in 1939.

In work whose impact reached far beyond the field of communication, J. B. Johnson discovered Johnson, or thermal, noise; Clinton Joseph Davisson, working with Lester Germer, demonstrated the wave nature of matter; Karl Guthe Jansky detected radio noise from our galaxy and so founded radio astronomy; and Arno A. Penzias and Robert W. Wilson discovered cosmic microwave background radiation and so verified the big-bang theory on the origin of the universe. Philip Anderson's work on solid state physics led to a Nobel prize. It is a testament to the quality of people and work at Bell Laboratories that eight of its people have been awarded the Nobel prize for work that was done there.

Work continues at Bell Laboratories, whose Murray Hill, New Jersey, facilities are shown in the photo. Much of the impact of its research will be felt in the future. Light-wave communication by means of fiber optics, the result of research in lasers and in optical fibers, has come to dominate long-distance transmission. But, as someone said about computers in the 1960s, after growing lustily, they are entering their infancy.

The Bell System is no more. AT&T Bell Laboratories serves solely the needs of AT&T. The research needs of the Bell operating companies, now independent of AT&T, are served by Bell Communication Research (Bellcore). Bellcore and AT&T Bell Labs share a common tradition of research in the art of communication.

is an increased emphasis on communication service and less emphasis on computers than immediately after divestiture.

With divestiture and competition, the communication industry has become fragmented, but the communication network and system must remain universal, and the fragmentation of the industry must not lead to a "tower of Babel." Clearly, research is even more important today to assure the future of telecommunication in the United States.

Ultimately, the effects of change from the vertically integrated Bell System to the present noncentralized means of supplying service will be best reflected in the relative progress in telecommunication in the United States compared to other advanced countries.

11

∙ ꞊ ꞊ ꞊ ꞊
∙ ꞊ ꞊ ꞊ ꞊

The Present and Future of Communication

Bell and Vail had a daring vision of universal telephone service—in their future. We have that today, and we have a lot of other communication services they never thought of. We may ask, what will we have tomorrow? Before you try to answer that question, I urge you to consider some of the unforeseen services that we have today—services that we take for granted when we know about them, or simply overlook because we haven't had occasion to use them. Let's dive into the woods of the present before trying to see trees in forests of tomorrow.

Touchtone Services

The touchtone keypad is more convenient than the old rotary dial, but that isn't its chief advantage. The dial produced on-off signals that didn't get beyond the central office equipment designed to

Visionaries have always enjoyed predicting the future. In this 1883 drawing, a young woman of the future does her social calling by air-car. The large number of wires accurately predicted the growth of communication but missed the use of multiplexing to conserve physical space.

receive them. Touchtone sends out pairs of tones that will go over the entire talking path when it has been set up—to any city, state, or nation.

Students are now able to register for their courses from their homes and dorm rooms simply by making a telephone call to the university's registration computer. The computer uses recorded or synthetic speech to prompt the student, and the student enters registration information using the touchtone keypad on the telephone. More than 50 major universities around the country are now using touchtone registration systems supplied by a number of vendors. A student usually completes the touchtone registration process in less than ten minutes.

Touchtone registration is only one example of the growing use of what is known as audiotext—a simple way to access data bases and to perform routine transactions using the familiar touchtone telephone and synthesized speech. In New York, Citibank customers use its CitiTouch^{SM} audiotext system to verify account balances, determine whether a check has cleared, and check interest rates.

Perhaps the commonest use of the touchtone pad is accessing from a distance messages recorded by one's home telephone answering machine. A Swedish telephone engineer told me years ago that he used the phone to turn on the heat in his vacation home before going there. The touchtone pad makes possible all sorts of remote operations and queries from any phone. We're just at the start of such services, large and small, institutional or personal. They creep up on us while we aren't watching; they become part of our lives before we realize that they're there.

Keeping in Touch Wherever You Are

We want to be able to reach some people in a hurry wherever they are—physicians, providers of emergency services, lawyers, and more occupations than you might think. Pagers tell the person paged to call in for a message. Some give a number to call back to.

Pagers are small, portable, battery-powered devices that beep in response to a radio signal addressed to a specific pager. The radio signal sent to the pager is at a frequency of about 1 GHz (1,000,000,000 Hz). All pagers in the area receive the signal, but the radio message includes a code so that only a specific pager responds. Newer pagers have a small liquid crystal display that gives the telephone number of the calling party.

With the aid of a communication satellite, some newer paging services are able to reach people in nearly the whole of the United States. The caller dials a toll-free telephone number, then enters the five-digit personal number of the person to be paged along with the caller's own telephone number. A message is sent by satellite to the local metropolitan area, where a paging message is broadcast. The pager "beeps" and displays the telephone number of the calling party.

It used to be that you were unreachable by your friends and colleagues once you were in the air traveling from city to city. Although you are still unreachable by them, they are not unreachable by you. Public telephones are now found in about one-third of the commercial fleet of airplanes in the United States, and that percentage might increase to all commercial airplanes, including overseas flights.

Telephony to airplanes is easy because airplanes fly high and can be seen line-of-sight from a few ground stations—about 80 ground stations cover the whole continental United States. The service in the United States is being offered by GTE and is called Airfone[R] service. The user makes the call as with any other telephone. The radio link to the nearest ground station is made, and the call is completed on the ground over GTE's long-distance network.

Extension to overseas flights might be made through satellite circuits, for one of the most important uses of satellites is to provide mobile service. Indeed, satellites are used in the United States to provide data channels from fleet headquarters to some thousands of trucks. Headquarters can interrogate the trucks and find their actual location through loran or other navigational equipment carried on the truck. But, it would be difficult to provide voice service to vehicles via satellite. Fortunately, another approach gives phone service to cars, and even to hand-held telephones, all over the country.

Cellular Mobile Telephony

Like long-distance telephony and transatlantic telephony, cellular mobile telephony is a grand service that took a lot of doing. In Los Angeles, where many people spend hours each day "parked" on freeways in traffic jams, car telephones are popular. Business people use them to stay in contact with their offices and customers. They are used to report traffic delays to the radio stations. Some

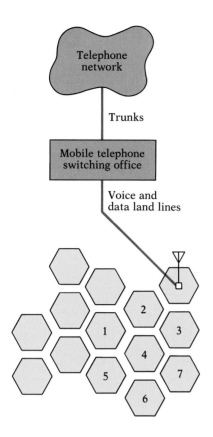

In a cellular mobile telephone system, a metropolitan area is divided into seven-cell clusters. Each cell is served by its own low-powered radio transmitter and antenna.

people use them to transmit facsimile from their car to the office. Mobile telephony is affordable and easily available thanks to the technology of cellular radio invented at Bell Laboratories and first used on a trial basis in Chicago in 1977. It came into commercial use in the early 1980s. Before cellular mobile telephone services, mobile telephony was accomplished using a single high-powered radio transmitter to cover a whole metropolitan area. Few channels were available, and service was poor, limited, and very costly. The solution was the use of many low-powered radio transmitters to cover the area. In this way, channels can be reused in different parts of the same overall metropolitan area.

In cellular telephony, the metropolitan area is divided into a number of clusters, each cluster consisting of seven cells. Each cell is served by its own low-powered radio transmitter, called a base station, and is allocated about 47 two-watt radio channels. A seven-cell cluster serves a total of 333 channels. The channels are reused in adjacent clusters of cells.

When a user leaves the jurisdiction of one cell, the user must be switched to one of the channels serving an adjacent cell. The handoff is accomplished automatically in a fraction of a second, and the user doesn't even notice. By monitoring the strength of the radio signal received in adjacent cells, the system can determine the optimum time to make the handoff. The information necessary to retune the mobile unit to a new channel is transmitted as a short burst of data over the voice channel.

The operation of the whole system is monitored and controlled at a mobile telephone switching office (MTSO). Trunks connect the MTSO to the local telephone network. Each base station serving each cell is connected to the MTSO by land lines that carry voice and data signals. A typical cell has a radius of from 6 to 12 miles. As the telephone traffic congestion increases, cells can be made smaller so that more cells are available to serve the increased number of users. Cellular telephony operates in the radio frequency range of 800 to 900 MHz.

Mobile telephones continuously monitor special radio channels that carry binary-encoded data. These so-called set-up channels notify the mobile units of incoming telephone calls and tell them the appropriate voice channel. The mobile units also request service for an outgoing call through the set-up channels. Each mobile phone has a unique telephone number and a serial number stored in it that can be queried by the system. Frequency modulation is used for the voice transmission, and each voice channel is allocated a bandwidth of 30 kHz.

CHAPTER 11

As the number of cellular mobile phones increases, it may be necessary to go to digital transmission and speech coding techniques to accommodate the traffic. Because digital transmission is more immune to interference, more channels can fit within the allocated frequency space. With digital transmission and linear predictive speech compression, the allocated UHF band should be able to handle seven times as much traffic. However, the large installed base of analog FM equipment could slow the introduction and growth of digital. Alternatively, more of the UHF frequency band could be allocated to the me-to-thee communication of mobile telephony and less to the they-to-us communication of television. Portable cellular telephones will continue to become smaller and less expensive. Dick Tracy's wrist telephone might indeed be closer than we think. Very low powered cellular systems might some day be a substitute for the pairs of copper wire found in the local loop and within office buildings. Cellular telephone service truly is a marvelous invention with an exciting future.

The Programmable Network

Through the advent of integrated electronics, through the lower cost of long-distance transmission, resulting principally from digital transmission and optical fibers, and through common-channel signaling, the telephone network has become more capable, flexible, and functional. Whereas in earlier days to change the system might have required extensive rewiring, or even the installation of new equipment, changes can now be provided simply by modifying the software, through programming the network, if you will. Flexibility and economy have made a number of advances possible.

Perhaps the most important of these advances is 800 service. I'm a customer of L. L. Bean. When I find an item I wish to purchase, I call an 800 number and reach their order department, which is staffed 24 hours a day. I give them my order and credit card number, and they ship promptly. About 30 percent of long distance calls in the United States are 800-number calls. Because the recipient pays for the call, it is free to the calling party.

The 800 call does not always go to the same location. When you make an 800 call, the 800 number must be converted to the appropriate routing number. Switching equipment sends the 800 number over the common channel signaling network to a data base that has stored the routing information for that particular

800 number. The routing information is then used to make the actual connection. The routing information may be different for different times of day, so that calls can be transferred to a part of the country where people are available to answer calls. Or, calls can be allocated to different business locations to ensure delivery from nearby centers, or to give a more nearly uniform workload.

911 service is important because one easily remembered number can be used for all sorts of emergencies. The newest and most sophisticated form of 911 service is called enhanced 911, or E911 for short. When a person picks up the telephone and dials 911, switching equipment automatically routes the call to the appropriate Public Safety Answering Point (PSAP), usually the local police dispatcher. This saves valuable time. The switching equipment has been programmed to perform this selective routing, thereby compensating for municipal boundaries that do not necessarily correspond to telephone company exchange boundaries.

The telephone number of the calling party is passed down the line to the PSAP. An automatic data call simultaneously goes out to a centrally located data base that contains the location corresponding to the calling party's telephone number, and this location is transmitted back to the PSAP. The number and location are displayed on a screen at the dispatcher's position at the PSAP while the dispatcher continues to speak to the person making the emergency call. The dispatcher can transfer the call to another emergency service if necessary, such as the fire department. If the calling party hangs up before giving the address and telephone number, the dispatcher has all the information necessary to call back.

Another emergency service allows those in precarious health to summon help merely by pressing a button on a tiny personal radio device. The radio then activates a telephone call to a central bureau that can summon help.

We have considered a few examples of well-established services provided through software, through programming the network. Other intriguing services are now being offered. Automatic caller indentification displays on a small special device the number from which the call is being made—before the phone is answered. Through call blocking one can reject calls from a particular number. Call forwarding is an older service; one can have all calls ring any designated telephone number for any designated period of time. Selective call forwarding is newer: only calls from designated telephone numbers are forwarded; or calls from designated numbers can be recorded and retrieved on calling in.

What services will the programmable network provide in the future? This depends on two things: ingenuity and consumer acceptance. People may not get all that they want, but they certainly will get only things they want and are willing to pay for.

The programmable network will furnish a large part of new communication in the future, and mostly a part that hasn't yet been proposed. Various other communication services have been proposed, things that don't sound at all like what we have been discussing—the services we have that neither Bell nor anyone else dreamed of. What about things someone has dreamed of?

The Comsole

Arthur C. Clarke is a very good writer of science fiction who was trained as a physicist to boot. In 1945, Clarke proposed satellite communication. For the future of communication, Clarke foresees a time when communication facilities will turn the whole world into Marshall McLuhan's global village. Clarke predicts that the people of his global village will have *comsoles*. Clarke's comsole is a universal communication terminal that provides two-way voice and picture communication, and that can retrieve any and all information known to the human race.

Can the comsole really be the future of communication? Or, is it just lumping together all sorts of past and present ideas? Mixing all sorts of ingredients together doesn't make the best possible dish—or the most digestible.

I've had some experience with the comsole, or attempted steps toward it. Let's start with information retrieval.

Information Retrieval

The LEXIS[R] system, offered by Mead Data Central, serves the legal profession. Mead's NEXIS[R] system contains the full text of a number of newspapers that can be searched by source or topic. MED-LINE, provided by the National Library of Medicine, serves physicians. The RLIN[R] system, operated by the Research Libraries Group, lists all the books held in member academic research libraries in the United States and arranges interlibrary loans. These are but a few of the professional data-base services. Two data-base services, the Source and CompuServe, are available to home computer users and provide electronic mail as well as various bulletin boards. Some of these systems link interested users worldwide. Over a thousand data bases are computer accessible. All it takes to use these services is a computer, a modem, and the appropriate telecommunication programs.

If you had the necessary equipment, could you find what you want to know through these extensive resources? Only if it's there; not everything is. Only if you are willing to pay; some of the services are expensive by personal standards. And, only if you master the different access and search procedures used by the different systems. We lack the standardization and good human engineering that could put even the *available* data at our fingertips.

Videotex

During the mid-1970s, researchers at the British Post Office invented an online data-base service for the home, which has since been called videotex. The service sent information to the home via the telephone connection. The user requested information using a keyboard terminal, and the home television set displayed the incoming text and crude graphics, in color. The service supplied such items as news, weather, sports scores, recipes, cinema schedules, and airline schedules. The data was organized into categories that were divided into finer and finer subcategories (called a tree structure) for easy access and retrieval by nonexpert users.

The British videotex system, called Prestel, was launched commercially in 1979. In 1978 the British Post Office predicted one million users by the end of 1981. The actual number of users was about 13,000.

While the British service was under way, AT&T in association with Knight-Ridder Newspapers ran trials and finally introduced a videotex service called Viewtron. The Viewtron data base contained a wide variety of information. However, people used the system mostly to post and send messages from one person to another. After a year of promotion, the service failed in the market, at a cost I estimate to be over $50 million. Videotex encountered a similar negative market response in Southern California, and the service offered there by Times-Mirror was likewise withdrawn.

Will it succeed? A number of banks developed their version of videotex: home banking from a terminal. Bills could be paid, funds transferred from one account to another, and a host of financial services accessed. The consumer did not care, and such home banking services as Chemical Bank's Pronto™ system failed too.

Perhaps if the terminals had dispensed cash, success would have been at hand! IBM and Sears have reportedly spent over $500 million developing their Prodigy^{SM} videotex service, introduced in 1988.

The only videotex system with mass-market status is the Teletel service offered in France by France Telecom. A terminal called Minitel is given away free to any telephone subscriber. Teletel provides directory assistance—the user does not need a local or national phonebook. The system can also send and receive electronic mail, and various vendors of products and services lease space in the system. In 1988 there were 4.3 million Minitel terminals in use in France, handling an average of 2.8 million calls per day—less than one daily call per terminal. At such a low usage compared to the telephone in the United States, does Teletel make money or is it a drain on the French Treasury?

The mass market for videotex-like information services seems doubtful, but there is a simpler form of information access that does appear successful. In Europe you can obtain mass-market,

Text and graphics are transmitted in several ways. Left: The British Oracle teletext service transmits several hundred pages, such as this index page, along with the conventional broadcast television signal. Right: IBM and Sears Prodigy^{SM} videotext service makes thousands of pages of information accessible over a telephone line and a personal computer. Advertisements, such as the one at the bottom of the weather map, help support the service.

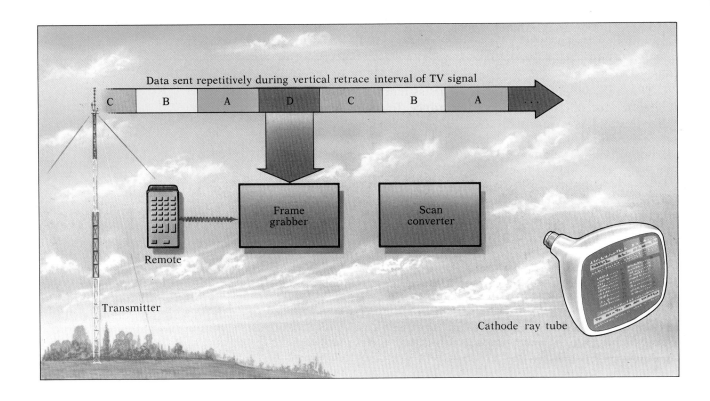

Data sent repetitively during vertical retrace interval of TV signal

C B A D C B A . . .

Remote

Frame grabber

Scan converter

Transmitter

Cathode ray tube

In teletext a broadcast television signal repeatedly transmits several graphics to a specially equipped television. After the user enters the desired frame number into the television's remote control, the frame is grabbed when it goes by and is converted for display on the TV screen.

general-interest information free via a system known as teletext. A teletext decoder built into the home TV set receives and displays pages of information consisting mostly of text with some simple color graphics. Teletext offers instant access to the weather, news headlines, sports scores, TV program listings, and other items of general interest. The teletext pages are transmitted along with the broadcast TV signal during the invisible vertical retrace interval, that is, the time allowed for the electron beam to return from the bottom of the TV screen back to the top to begin scanning another picture frame. Some hundreds of pages are sent repeatedly in sequence. The viewer keys a request to display a particular page by entering the number of the desired page on the same remote control device that is used to change channels. When the requested page arrives at the TV set, it is grabbed, stored, and displayed on the TV screen.

Teletext is a reasonable success in Europe. About one-third of TV households in Great Britain have TV sets equipped for teletext, and nearly every TV set presently sold has a teletext capability.

Pictures Yet!

Some information is at our fingertips, but what we have falls far short of what Clarke expects of the comsole of McLuhan's global village. What about the two-way pictures with communication that the comsole provides? If we don't yet have them, it isn't for want of trying.

The idea of two-way video communication—the video telephone or picturephone—is old *outside* of science fiction. A two-way videophone was exhibited at the Chicago World's Fair in the early 1930s, before network television. Coaxial cable linked together public videotelephone centers in four cities in Germany from 1935 to 1938. AT&T first demonstrated its picturephone system for use in homes and businesses at the New York World's Fair in 1964. Later that year, the company initiated limited commercial service between public centers in three cities. After conducting trials of the service within various businesses, AT&T introduced picturephone to Pittsburgh in 1970 and to Chicago in 1971. The service then cost $160 for the first 30 minutes per month, and 25 cents for each additional minute. In 1971, a West Coast think-tank predicted over 2 million picturephone sets in use by 1985. How wrong they were!

Even in 1929 futurists foresaw the humorous, if negative, implications of combining the telephone with television.

This 1969 picturephone marketed by AT&T was a sleek, sophisticated system offering two-way video communication. However, most consumers did not have a strong desire to see the person they were speaking to, and the picturephone was withdrawn.

AT&T's video teleconferencing service, beginning in the early 1970s, had much logical appeal as an alternative to business travel to attend meetings. The service was available between specially equipped rooms in New York City, Washington, D.C., Los Angeles, Chicago, and San Francisco. Color was added as well as special graphics features, but the usage remained low. Some researchers had predicted that teleconferencing could take the place of from 20 to 85 percent of all meetings involving travel. But like picturephone, the service failed and faded away. Two-way teleconferencing facilities still link the scattered offices of some large companies, but teleconferencing has never really taken off. A study I performed about 10 years ago indicated that only 4 percent of all meetings involving groups of people are good candidates for teleconferencing.

Will optical fiber, broadband video-switching systems, and video technology make picturephone so cheap that we will all want it? Will newer video teleconferencing technology create a "video window" or even a "telepresence" so realistic that video teleconferencing will become desirable? Some believe that because we are a video society, video must also be the way of the future in two-way communication. I doubt it, and indeed, many consumers are so against the picturephone that they would pay extra not to have it!

The Future of Communication

Progress toward the future is often evolutionary rather than revolutionary. But continual improvements in technology can only go so far, and then it is time for a totally new, improved replacement. Evolution stops and revolution occurs.

But revolutions must succeed in the marketplace. Economics, regulatory policy, and consumer response can determine the future of communication. And, considerable risk and financial investment are involved in introducing new products and services into the market. I think of Alexander Graham Bell, his vision and invention of the telephone, and the risks taken by his financial backers in turning a dream into reality and success. Without such dreams the future will be bleak indeed.

Through large-scale integrated circuits and computers, through digital communication by optical fibers, the communication networks of tomorrow will in principle do about anything that we are smart enough to think of, and smart enough to make them do, just by changing software. Perhaps we will be able to provide inexpensive, truly useful, easily and widely used information services that will tell people things they really want to know—in recreation, on the job, in coping with the problems of life. Or, perhaps some enhancement of the telephone will enable us to escape from life into the thrilling video games, or to obtain TV shows and movies on request by wire, or to choose the camera shot while watching a TV sporting event.

Granted a human society with a thriving technology, all the dreams of vast amounts of information at our beck and call, even unto Arthur Clarke's comsole and global village, may some day be ours. We may one day travel for pleasure and communicate to work.

Whatever we may say of the future, it is open to us. That is, if we are knowledgeable enough to act, and if we leave ourselves free to act.

Further Readings

BOETTINGER, H. M., *The Telephone Book: Watson, Vail and American Life, 1876–1976.* Riverwood Publishers, Ltd., Croton-on-Hudson, 1976.
The story of the telephone for the general reader, as well as for those with a special interest in technology. Richly illustrated.

BROOKS, JOHN, *Telephone.* Harper and Row, New York, 1975.
The history of AT&T and the Bell System, from its earliest years to the coming of competition.

CLARKE, ARTHUR C., *Voice Across the Sea.* William Luscombe, London, 1974.
The story of early cables and communication satellites.

COLL, STEVE, *The Deal of the Century: The Breakup of AT&T.* Simon and Schuster, Inc., New York, 1986.
A blow-by-blow account of the antitrust suit, from two years before its initiation on January 2, 1974, to the ultimate resolution on January 2, 1982. Fascinating and enlightening, especially the nontechnical side.

DE SOLA POOL, ITHIEL (editor), *The Social Impact of the Telephone.* The MIT Press, Cambridge, Massachusetts, 1977.
A collection of essays by social scientists and engineers focusing on the evolution of the telephone into an essential means of communication.

Engineering and Operations in the Bell System. Bell Telephone Laboratories, Inc., 1977.
A thorough and understandable description of telephone engineering.

FAGEN, M. D. (editor), *A History of Engineering and Science in the Bell System: The Early Years (1875–1925).* Bell Telephone Laboratories, Inc., Holmdel, New Jersey, 1975.
A detailed examination of the innovations of the Bell System, with sketches and original photographs, tracing the development of telephone technology.

INOSE, HIROSHI, *An Introduction to Digital Integrated Communications Systems.* The University of Tokyo Press, 1979. An authoritative explanation by a pioneer in the field of digital transmission, switching, and their applications to communication.

INOSE, HIROSHI, and JOHN R. PIERCE, *Information Technology and Civilization.* W. H. Freeman and Company, New York, 1984. A discussion of the current and future social and cultural ramifications of information technology.

LUCKY, ROBERT W., *Silicon Dreams: Information, Man, and Machine.* St. Martin's Press, New York, 1989. A wonderful book covering all aspects of information, for the eager layperson or the technological expert.

MABON, PRESCOTT C., *Mission Communications: The Story of Bell Laboratories.* Bell Telephone Laboratories, Holmdel, New Jersey, 1975. An illustrated history of the people and the contributions of Bell Laboratories for all readers.

NOLL, A. MICHAEL, *Introduction to Telecommunication Electronics.* Artech House, Inc., Norwood, Massachusetts, 1988. Hand-drawn figures explain the workings of electronics for the general reader in a friendly fashion.

PIERCE, JOHN R., *An Introduction to Information Theory: Symbols, Signals and Noise.* Dover Publications, New York, 1980. A broad discussion of communications.

SHANNON, CLAUDE E., and WARREN WEAVER, *The Mathematical Theory of Communication.* The University of Illinois Press, Urbana, 1963. Shannon's now classic first paper on communication, with a discussion by Warren Weaver of more recent applications of Shannon's theories.

VON AUW, ALVIN, *Heritage and Destiny.* Praeger Publishers, New York, 1983. Personal reflections on the Bell breakup by a retired AT&T executive.

Sources of Illustrations

All drawings by Vantage Art unless otherwise identified

Facing page 1
Library of Congress, Grosvenor Collection

Page 2
Telefocus; a British Telecom photograph

Page 3
Snark/Art Resource

Page 4
Bellcore

Page 5
Drawing by Tom Moore

Page 9
Library of Congress, Grosvenor Collection

Page 10
Jeff MacWright

Page 12
left, Telephone Pioneer Communications Museum, San Francisco
right, The Telecom Technology Showcase (British Telecom Museum), London

Page 14
NASA

Page 15
Bellcore

Page 16
The Telecom Technology Showcase (British Telecom Museum), London

Page 18
The Bettmann Archive

Page 19
The Telecom Technology Showcase (British Telecom Museum), London

Page 20
Smithsonian Institution Photo No. 74-2444

Page 21
HPL/Bettman Archive

Page 29
Adapted from H. Fletcher, *Speech and Hearing in Communications*. New York, Van Nostrand, 1953.

Page 30
AT&T Archives

Page 33
Drawing by Tom Moore

Page 37
AT&T Archives

Page 38
AT&T Archives

Page 39
AT&T Archives

Page 40
AT&T Archives

Page 42
Art Resource

Pages 52 and 53
Drawing by Tom Moore

Page 54
AT&T Archives

Page 56
Adapted from C. Shannon and W. Weaver, *The Mathematical Theory of Communication*. University of Illinois Press, 1949.

Page 62
Jet Propulsion Laboratory

Page 63
AT&T Archives

Page 66
Telefocus; a British Telecom photograph

Page 78
Canadian National

Page 84
NKF Kabel B. V.

Page 86
Museum of the City of New York

Page 88
Telefocus; a British Telecom photograph

Page 89
Wide World Photos

Page 93
From "Communication Terminals," by E. Kretzmer. Copyright © 1972 by Scientific American, Inc. All rights reserved.

Page 94
AT&T Archives

Page 99
From "Communication Terminals," by E. Kretzmer. Copyright © 1972 by Scientific American, Inc. All rights reserved.

Page 101
Alcatel

Page 102
Intelsat

Page 108
Stephen Dalton/NHPA

Page 112
The Telecom Technology showcase (British Telecom Museum), London

Page 113
Drawing by Tom Moore

Page 117
Telefocus; a British Telecom photograph

Page 118
Martin Dohran/Science Source/Photo Researchers

Page 121
Telefocus; a British Telecom photograph

Page 128
Telefocus; a British Telecom photograph

Page 129
NKF Kabel B. V.

Page 130
From Suzanne R. Nagel, "Optic Fiber—The Expanding Medium." *IEEE Communications Magazine*, Vol. 25, No. 4, page 36, April, 1987, © 1987 IEEE.

Page 133
Telefocus; a British Telecom photograph

Page 134
The Telecom Technology Showcase (British Telecom Museum), London

Page 138
Marconi Company Ltd.

Page 139
Smithsonian Institution Photo No. 72,132

Page 141
G. E. Hall of History

Page 142
The Granger Collection

Page 144
Varian Associates, Inc.

Page 147
left, AT&T Archives
right, AT&T Archives

Page 148
The Telecom Technology Showcase (British Telecom Museum), London

Page 152
Adapted from Thomas M. Frederiksen, *Intuitive CMOS Electronics.* McGraw-Hill Book Company, 1989.

Page 153
left, Motorola, Inc.
right, Hewlett Packard Company

Page 156
Adapted from Joseph C. Palais, *Fiberoptic Communications*, Second Edition. Prentice Hall, 1988.

Page 158
Hughes Aircraft Company

Page 160
The Telecom Technology Showcase (British Telecom Museum), London

Page 163
Telephone Pioneer Communications Museum, San Francisco

Page 170
The Telecom Technology Showcase (British Telecom Museum), London

Page 171
From *Engineering and Science in the Bell System.* AT&T Archives

Page 175
AT&T Archives

Page 178
AT&T Archives

Page 185
AT&T Archives

Page 190
The Bettman Archive

Page 195
James Kilkelly

Page 196
AT&T Archives

Page 198
Jet Propulsion Laboratory

Page 199
Telefocus; a British Telecom photograph

Page 200
AT&T Archives

Page 201
Alcatel

Page 206
Alcatel

Page 207
Alcatel

Page 210
The Telecom Technology Showcase (British Telecom Museum), London

Page 215
Telefocus; a British Telecom photograph

Page 217
Bell Canada

Page 218
Alcatel

Page 219
AT&T Archives

Page 220
AT&T Archives

Page 222
AT&T Archives

Page 224
Ann Ronan Picture Library

Page 233
left, Oracle
right, Prodigy

Page 234
Drawing by Tom Moore

Page 235
From the Smithsonian Institution Traveling Exhibition Service (SITES) traveling exhibition, "Yesterday's Tomorrows: Past Visions of the American Future." Photo by Joe Goulait, Smithsonian Institution.

Page 236
AT&T Archives

Index

Other books in the Scientific American Library Series